OTHER BOOKS ON STATISTICS AND MATHEMATICS

Descriptive brochure from Charles Griffin & Co. Ltd.

PATTERNS AND CONFIGURATIONS
IN
FINITE SPACES

S. VAJDA
Dr. Phil. (Vienna), F.S.S.
Professor of Operational Research
University of Birmingham

BEING NUMBER TWENTY-TWO OF
GRIFFIN'S STATISTICAL
MONOGRAPHS & COURSES
EDITED BY
ALAN STUART, D.Sc.(Econ.)

1967
HAFNER PUBLISHING COMPANY
NEW YORK

Copyright © 1967

CHARLES GRIFFIN & COMPANY LIMITED
42 DRURY LANE, LONDON, W.C.2

SBN: 85264 025 0

First published . . . 1967

SET BY E. W. C. WILKINS & ASSOCIATES LTD, LONDON
PRINTED IN GREAT BRITAIN BY LATIMER TREND & CO LTD, WHITSTABLE

PREFACE

The topics with which the two companion volumes

Patterns and Configurations in Finite Spaces
and
The Mathematics of Experimental Design : incomplete
block designs and Latin squares

are concerned have a long history, but the main stimulus to new developments has come from the interest of statisticians in the efficient design of experiments. The origin of some of the patterns can be traced to recreational mathematics, or to problems in pure mathematics, particularly in number theory.

The statistical aspect of most of the subjects mentioned has been dealt with in many excellent textbooks, together with their analysis which leads to inferences about the effectiveness of treatments or other choices, the outcome of which is subject to stochastic variation. No such analysis is described in the present books. They contain rather the combinatorial aspects of the construction of designs, without regard to their practical application or indeed other usefulness.

In *Patterns and Configurations in Finite Spaces*, the first chapter contains the algebra which is needed in the subsequent pages; it is, of course, not a course in algebra, but only a review of certain parts of it.

Chapters II and III contain the fundamentals of finite geometry, two-dimensional and more-dimensional geometry being treated in separate chapters. In Chapter IV we reach configurations.

In *The Mathematics of Experimental Design* we start with a short review of algebraic facts, and then deal fully with some special cases of configurations treated in the first-mentioned book. Thus, balanced incomplete block designs are described in Chapter II, and partially balanced incomplete block designs in Chapters IV and V, the last chapter being concerned with a special case of designs introduced in the former, namely those of two associate classes.

Chapter III is devoted to orthogonal arrays and to a special case of these, namely Latin squares and their orthogonal sets. It contains an account of the most recent results in this field, including the proof of the falsity of a hypothesis by Euler, which was long considered to be plausible.

The bibliographies mention only those papers which are referred to in the text of the respective book. A complete bibliography of the subject would be too extensive.

The exercises are meant to test the reader's understanding, and to serve the teacher of these topics. Both the reader and the teacher will welcome the inclusion of solutions to the exercises. Examples are also found throughout both books.

Birmingham, 1967

S. VAJDA

CONTENTS

CHAPTER I

ABSTRACT ALGEBRA

In the construction of the various patterns and configurations which will be our main concern in this book, we shall be aided by a knowledge of some algebraic concepts, and this chapter is devoted to their study, to the extent needed to give a proper background to the sequel.

Finite groups

A finite group is a set of a finite number of distinct elements, say $a_1, \ldots, a_s$, with a rule of operation for combining any two of them in a given order. The operation satisfies the following rules:

(1) If both a_i and a_j are elements of the group, then their combination, denoted by $a_i a_j$, and called their *product*, is also an element of the group.

(2) If a_i, a_j, and a_k are three elements of the group, then $(a_i a_j) a_k = a_i (a_j a_k)$ (associative law).

(3) The group contains an element, I, say, called the *identity element*, such that for every element a_i of the group we have $a_i I = a_i$.

(4) For every element a_i of the group there exists an element, a_j say, such that $a_i a_j = I$. We write $a_j = a_i^{-1}$.

As a consequence of these rules we can show that

(i)   if $a_i a_j = I$, then $a_j a_i = I$;

(ii) $I a_i = a_i$ for all a_i (the definition of I implies only $a_i I = a_i$);

(iii)   the identity element of the group is unique;

(iv)   the inverse of an element is unique.

Proofs

(i)   Let $a_i a_j = I$, and $a_j a_k = I$ (such a_j and a_k exist for every a_i).
Then $a_i = a_i I = a_i (a_j a_k) = (a_i a_j) a_k = I a_k$.   .   .   .   . (i_0)
and hence $a_j a_i = a_j (I a_k) = (a_j I) a_k = a_j a_k = I$.

(ii)   Let $a_j a_k = I$, and $a_i a_j = I$. Then
$$I a_i = (II) a_i = (I(a_j a_k)) a_i = (I(a_k a_j)) a_i \quad \text{by (i)}.$$

Continuing, we have $(Ia_k)(a_j a_i) = (Ia_k)I$ by (i) and, from (i_0), $(Ia_k)I = a_i I = a_i$.

(iii) Let I_1 and I_2 be identity elements. Then

$$I_1 = I_1 I_2 = I_2 I_1 = I_2,$$

i.e. there is only one identity element in the group.

(iv) Let $a_i a_j = I$, and $a_i a_k = I$. Then

$$a_k = a_k I = a_k(a_i a_j) = (a_k a_i)a_j = Ia_j \quad \text{by (i)},$$

and hence $a_k = a_j$.

We see also that for any pair of elements, a_i and a_k, say, there exists an a_j such that $a_i a_j = a_k$. It is the element $a_i^{-1} a_k$ and it is, of course, unique.

It should be noticed that we have not assumed that the operation of combining is commutative, i.e. that $a_i a_j = a_j a_i$ for all i,j, although this is so if either element is I. If the combination is always commutative, then the group is called *Abelian*.

As we have seen, the identity element commutes with every element of the group, whether the group is Abelian or not. An element which commutes with every other element is called *self-conjugate, invariant*, or *normal*.

We now quote some theorems about groups which we shall use in later chapters.

Theorem 1.1 A group of order p^n, where p is a prime and n is a positive integer, has at least one invariant element other than the identity ([15]; p. 68).

The number of elements in a group is called its *order*. Consider the powers of an element a_i, namely $a_i a_i = a_i^2, \dots, a_i^t, \dots$ Since the number of all elements is finite, there must be a repetition of elements if we extend the series sufficiently far. Let $a_i^s = a_i^t$, and $s > t$. We have then $a_i^{s-t} a_i^t = a_i^s = a_i^t$, and

$$(a_i^{s-t} a_i^t)(a_i^t)^{-1} = a_i^{s-t}(a_i^t)(a_i^t)^{-1} = a_i^t(a_i^t)^{-1},$$

so that $a_i^{s-t} = I$.

The smallest integer m, such that $a_i^m = I$, is called the *order* of a_i. If $a_i^j = I$, then j is a multiple of m. For, if $j = am + b$ (a, b integers and $0 \leqslant b < m$), then $a_i^j = (a_i^m)^a . a_i^b$, i.e. $a_i^b = I$, and hence $b = 0$.

An element and all its powers form an Abelian group, whose order equals that of the element. Such a group is called *cyclic*, and all groups whose order is a prime are such cyclic groups.

We shall prove presently that the order of every element of a group is a factor of the order of the group.

First, we introduce the concept of a *sub-group*. This is a set of elements of a group which form a group themselves. The whole group, and the identity element by itself, are two such sub-groups. The other sub-groups (if any) are called "proper sub-groups".

Consider a proper sub-group of order m, consisting of $a_1 = I$, a_2, ..., a_m. Take an element outside this sub-group, say b, and form $ba_1, ..., ba_m$. They are all different, and different from the elements of the sub-group. If we have then accounted for all the elements of the group, then its order is $2m$. Otherwise we take yet another element, say c, not yet considered, and form $ca_1, ..., ca_m$. They are again different from one another and from any element already considered, for if $ba_s = ca_r$, then c would be equal to $ba_s a_r^{-1} = ba_t$, say.

Continuing until all elements of the group have been accounted for, we see that the order of the group is a multiple of the order of the sub-group. The cyclic group formed by the powers of an element is also a sub-group, and hence the order of any element divides the order of the group.

Theorem 1.2 (theorem of Frobenius) Let G be a group of order g and let n be a factor of g. Then the number of elements of G, including the identity, whose orders are factors of n, is a multiple of n ([15], p. 92).

If $g = 4t+2$, then 2 is one of the factors of g. Hence the number of elements with orders 1 or 2 is at least 2. This includes the identity, and it follows that there must be at least one element of even order in the group.

Another factor of $4t+2$ is $2t+1$, and all its factors are odd. Therefore the number of elements of odd order is a multiple of $2t+1$. But it cannot be $4t+2$, because elements of even order also exist in the group. Hence both the number of elements of odd order, and of even order, is precisely $2t+1$.

Two groups are called *isomorphic* if each element of one can be made to correspond to an element of the other in such a way that the product of two elements from one group corresponds to the product of the corresponding elements of the other. If the correspondence is one-to-one, then the groups must be of equal order, and the two identity elements will correspond to one another, because each power of I is again I. The isomorphism is then called *simple*.

A simple isomorphism of a group with itself is called an *automorphism*.

H.B. Mann, in [54], calls an automorphism in which the products of two corresponding elements are all different, a "complete mapping". For instance, in a group of odd order the correspondence of each element to itself is complete, because all squares of the elements of the group are different. They form again the same group.

A complete mapping of a group of order $4t+2$ cannot exist. Let us assume it did, mapping x_i into y_i. Then all $4t+2$ elements of the group are of the form $x_i y_i$. Let there be a such elements where both x_i and y_i are of odd order, b elements where x_i is of odd and y_i is of even order, and c elements where x_i is of even and y_i is of odd order. Then $a+b = a+c = 2t+1$. Also, there are then $b+c$ elements of even order (those where either x_i or y_i, but not both, are of even order), and hence $b+c = 2t+1$. But then $2a+b+c = 4t+2$, $2a = 2t+1$, which is impossible.

We also need the concept of the direct product of two groups, with elements a_i $(i = 1, ..., n_1)$ and b_j $(j = 1, ..., n_2)$ respectively. It consists of all products $a_i b_j$ and it is assumed that $a_i b_j = b_j a_i$. The order of the direct group is clearly the product of the orders of the two original groups. The direct product of two Abelian groups is again Abelian.

Theorem 1.3   Every group of order $p_1^{m_1} ... p_k^{m_k}$ where the p_i are different primes, has at least one sub-group of order $p_i^{m_i}$ for all i (these sub-groups are called *Sylow sub-groups*), and this sub-group contains at least one representative of each class of elements of order p_i in the original group ([15], pp. 58 ff.).

In this context, "class" is defined as a set of all elements such that any two of them, say x_1 and x_2, can be derived from one another by $x_1 = y^{-1} x_2 y$, where y is a member of the group.

Since there cannot be more than $p_i^{m_i} - 1$ elements of order p_i in a group of order $p_i^{m_i}$ (because the group contains also I), there cannot be more than $p_i^{m_i} - 1$ classes of elements of order p_i in the group.

The following theorems refer to Abelian groups:

Theorem 1.4   A non-cyclic Abelian group of prime power order p^m is the direct product of cyclic sub-groups, no two of which have any element in common apart from I ([15], pp. 66, 98).

If the orders of the cyclic sub-groups are $p^{m_1}, ..., p^{m_k}$ $(m = m_1 + ... + m_k)$ then we say that the group is of type $(m_1, ..., m_k)$. We shall in particular be concerned with groups of type $(1, ..., 1)$. Clearly, any two Abelian groups of this type and of equal order are isomorphic.

Theorem 1.5 If the elements of the highest order in an Abelian group have order k, then any other order, say t, is a factor of k.

Proof: If this were not so, then there would exist some prime p such that $k = p^e r$, and $t = p^f r'$, where r and r' do not contain p as a factor, and $f > e$. Let a be an element of order k, and b an element of order t. Consider the element $a^{p^e} b^{r'} = c$, say.

We have $c^{p^f r} = I$, and we can prove that $p^f r$ is, in fact, the order of c. This is seen as follows:

Let the order of c be m, thus $(a^{p^e} b^{r'})^m = I$. Because the group is Abelian, the left-hand side can be written $(a^{p^e})^m (b^{r'})^m$. From $a^{p^e r} = I$ we have $b^{r'rm} = I$, or $r'rm = O(\mathrm{mod}\ p^f r')$; hence $rm = O(\mathrm{mod}\ p^f)$ and, because p and r are relatively prime, $m = O(\mathrm{mod}\ p^f)$.

Similarly, from $b^{p^f r'} = I$, we have $a^{p^e p^f m} = I$, then $p^e p^f m = O(\mathrm{mod}\ p^e r)$; hence $p^f m = O(\mathrm{mod}\ r)$ and $m = O(\mathrm{mod}\ r)$.

From $m = O(\mathrm{mod}\ p^f)$ and $m = O(\mathrm{mod}\ r)$ we have, since p and r are relatively prime, that $m = O(\mathrm{mod}\ p^f r)$. But $c^{p^f r} = I$, and therefore $m = p^f r$. However, $p^f r > p^e r = k$, and thus it appears that we have found an element, namely c, of higher order than k, the highest order of all elements in the group. This contradiction can only be resolved by assuming that all orders must be factors of the highest order of all elements.

Permutation groups

The permutations of m symbols form a group of order $m!$, and m is called the *degree* of any group of permutations of m symbols. Since every group of permutations of degree m is a sub-group of the group of all permutations of m symbols, i.e. of the "symmetric" group of m symbols, the order of any such permutation group must be a factor of $m!$

It is convenient to describe a permutation by a succession of cycles. Each cycle, written $(abc... d)$ indicates that the permutation changes a into b, b into c, ..., and finally d again into a. For instance, the permutation $(12)(34)$ exchanges 1 and 2, and also 3 and 4. For the identity permutation which leaves all symbols unchanged, we write E.

Example

It is easily verified that the permutations E, (12) (34), (13) (24), (14) (23) form a group. This group is Abelian, but not cyclic.

The importance of permutation groups is due to the fact that every group of finite order is simply isomorphic with a permutation group. This can be shown as follows:

Let the elements of a group be ordered in some way, say a_1, ..., a_n. Premultiplication by an element a_i other than I produces n distinct elements, $a_i a_1$, ..., $a_i a_n$, hence these elements form a permutation of the elements a_1, ..., a_n.

If we let an element of the group correspond to the permutation which it thus produces, then the permutations form a group which is simply isomorphic to the original group. We could equally well have used postmultiplication, and would have obtained another permutation group, unless the original group was Abelian.

The degree of the permutation group thus obtained is equal to its order. If we take an element a_j of the original group, then we obtain in the way described a permutation which changes it into a_k, namely the permutation corresponding to the element $a_k a_j^{-1}$. This means that in the permutation group obtained we can find for any two given symbols a_j, a_k a permutation which replaces one by the other. Such a permutation group is called *transitive*, and a transitive group whose degree equals its order is called *regular*.

In a regular group of order (and degree) n there are n different permutations which replace a_i, respectively, by a_1, ..., a_n. These are all the permutations of the group. Let T_i be the permutation which replaces a_i by itself; it must be E: the identity permutation is the only one which leaves any symbol unaltered. Hence, if we write the permutations as sequences of cycles, then all the cycles in the same permutation must have the same length, because otherwise one of its powers different from E would leave at least one symbol unaltered.

Algebras A[s]

An algebra $A[s]$ is a set of $s(>1)$ distinct elements, which satisfy certain rules of two types of combinations. Because of the similarity of these operations to addition and multiplication in ordinary algebra, we give them the same names, and denote them by a plus sign and a dot respectively. The dot will be omitted when this is possible without any risk of misunderstanding.

The two types of combination are defined as follows: with regard to addition, the elements form a group. The identity element will be denoted by 0, and the inverse of **a** by −**a**. With regard to multiplication, all elements except 0 form a group. The identity element will be denoted by 1, and the inverse of **a** by a^{-1}.

Further, we have a rule which establishes a connection between the two types of combination:

For any three elements a_i, a_j, and a_k, we have

$$a_i(a_j + a_k) = a_i a_j + a_i a_k \quad \text{(distributive law)}.$$

Note that we do not assume that

$$(a_j + a_k)a_i = a_j a_i + a_k a_i$$

is also necessarily true. But when multiplication is commutative, then clearly the second distributive law follows from the first.

On the other hand, it has been proved by Wedderburn, and by Dickson (see [20], p. 379) that if both distributive laws hold, then the commutativity of multiplication follows.

Applying the distributive law when one of the elements is 0, we have for any element a_i

$$a_i.0 = a_i(0 + 0) = a_i\,0 + a_i.0,$$

so that $a_i.0 = 0$. That $0.a_i$ is also 0 can be seen as follows: If it were not 0, then it would have an inverse under multiplication, say a_j. Then $(0.a_i).a_j = 1$. Now

$$0.a_i = (0.0).a_i = 0.(0.a_i), \quad \text{and hence}$$

$$1 = (0.a_i).a_j = (0.(0.a_i)).a_j = 0.((0.a_i).a_j) = 0.1 = 0.$$

But this is impossible, because it would mean that for any element a_i we have $a_i = a_i.1 = a_i.0 = 0$, and there would be only one single element in the algebra.

If the product of two elements equals 0, then at least one of the factors must be 0.

Proof: If $a_i.a_j = 0$, and $a_j \neq 0$, then

$$a_i = a_i.(a_j.a_j^{-1}) = (a_i.a_j)a_j^{-1} = 0.a_j^{-1} = 0.$$

Thus the identity element under addition has all the properties associated with the symbol 0 in ordinary algebra.

The elements $1, 1+1, (1+1)+1, \ldots$ of $A[s]$ are called *integers*. As the number of elements added increases, there must occur a repetition of elements. It can be proved (the proof is analogous to that in group theory), that after a finite number of additions we reach 0. Let this be the first time when p terms have been added. Then there exist exactly p integers in the algebra. We denote them by $b_1 = 1, b_2, \ldots, b_p$. Their addition is equivalent to that of integers in ordinary algebra, and reduction modulo p. Their multiplication is equivalent to multiplication in ordinary algebra, and again reduction modulo p. This follows, because $(1+1+\ldots+1).(1+\ldots+1)$, with u terms in the first and v terms in the second bracket, equals $(1+1+\ldots+1).1+\ldots+(1+1+\ldots+1).1$ $= 1.1+\ldots+1.1$ with $u.v$ terms. Now $u.v$ (mod p) equals one of the numbers $1, 2, \ldots, p$, and hence the product considered equals one of the integers. It follows that the set of integers itself forms an algebra, $A[p]$.

The number of integers, p, must be a prime, because if $p = u.v$, and $b_u \neq 0$, then $b_u.b_v = b_p = 0$, so that $b_v = 0$. But this is a contradiction, unless $v = p$. In what follows, we shall denote the integers simply by $0, 1, \ldots, p-1$.

If x is a non-zero integer, then its inverse under addition, and its inverse under multiplication, are also integers.

We have seen that the number of integers in an algebra $A[s]$ is a prime number, say p. We show now that the *order* of the algebra, namely s, is a power of the prime p.

Let w_1 be a non-zero element of the algebra. Take all elements $w_1 q_1, \ldots, w_1 q_p$, where q_i are the p integers in the algebra. When these are all the elements, then the statement is obviously true. Otherwise take one of the remaining elements, say w_2, and form all the elements $w_1 a_1 + w_2 a_2$, where a_1 and a_2 are independently chosen from the integers. These p^2 elements are all distinct, for if

$$w_1 a_1 + w_2 a_2 = w_1 b_1 + w_2 b_2$$

then

$$w_1 a_1 + (-w_1 b_1) = w_2 b_2 + (-w_2 a_2)$$

and using the distributive law, this can be written

$$w_1(a_1 - b_1) = w_2(b_2 - a_2).$$

Thus, if $b_2 - a_2 = 0$, then $b_1 - a_1 = 0$ as well, because $w_1 \neq 0$. On the other hand, if $b_2 \neq a_2$, then

$$w_2 = w_1(a_1 - b_1).(b_2 - a_2)^{-1},$$

i.e. w_2 would have appeared already among the multiples of w_1.

If there is a further element, say w_3, then we form

$$w_1 a_1 + w_2 a_2 + w_3 a_3$$

by independent choices of the a_i from the integers, and it can again be proved that the resulting elements are all different. We continue in this way until all elements are accounted for. We have thus proved:

Theorem 1.6   The number of elements in an algebra $A[s]$ is a prime power.

Let x be an element of an algebra $A[s]$ and compute $x + x + \ldots + x$, a sum of p terms. This sum equals $x.(1 + \ldots + 1) = 0$, so that the order of the element x, as an element of the additive group, is a factor of p. But p is a prime, so that all elements except 0 must have precisely the order p.

We prove now that addition in an algebra is commutative.

The additive group has order $s = p^n$, and therefore it contains an invariant element other than the identity (see Theorem 1.1). Let such an element be y, so that $x + y = y + x$ for every element x.

Let z be a non-zero element and determine x' so that $x + z = z + x'$. We want to prove that $x = x'$.

Multiplying $x + z = z + x'$ on the left by $y.z^{-1}$, and using the distributive and associative laws, we have

$$y.z^{-1}.x + y.z^{-1}.z = y.z^{-1}.z + y.z^{-1}.x', \quad \text{i.e.}$$

$$y.z^{-1}.x + y = y + y.z^{-1}.x'.$$

But y is an invariant element, so that $y.z^{-1}.x + y = y.z^{-1}.x' + y$, and hence $x = x'$. Thus the commutativity of addition within an algebra $A[s]$ is proved.

It follows that the additive group within an algebra $A[s]$ is of type $(1, 1, \ldots, 1)$, because, as we have seen, the order of all non-zero elements is p.

Two algebras $A_1[s]$ and $A_2[s]$ are said to be *isomorphic* if the elements can be put into a one-to-one correspondence so that the sum and product of any pair of elements of the one correspond to the sum, and the product, respectively, of the corresponding pair of elements in the other. Clearly their additive groups of order s as well as their multiplicative groups of order $s-1$ are then also isomorphic. The first fact is in any case true of any two algebras with the same number of elements, because the additive group is of type $(1,\ldots,1)$. On the other hand, the isomorphism of the multiplicative groups is not a consequence of the equality of those numbers. It can be shown that this latter isomorphism is also sufficient for isomorphisms of two algebras $A_1[s]$ and $A_2[s]$ ([15], p. 402).

It is known that there exist four distinct, not isomorphic algebras $A[9]$. This is the algebra of smallest order for which no isomorphic alternatives exist. There are two distinct $A[81]$. All $A[p]$ are, of course, unique if p is a prime. Other values of s for which $A[s]$ has been shown to be unique are, for instance, some powers of 2, and 243.

Example

We quote one $A[9]$ from [105], also given in [15], p. 410. The elements are of the form $a + ib$, $a, b = 0, 1, 2 \pmod 3$. Addition is defined as for complex numbers, i being identified with $\sqrt{-1}$. For multiplication we use the distributive law and interpret $(a + ib).i$ as $-ai + b$, while $i.(a + ib) = ia - b$, as for complex numbers. This gives the following table of multiplication:

	0	1	2	i	$2i$	$1+i$	$1+2i$	$2+i$	$2+2i$
0	0	0	0	0	0	0	0	0	0
1	0	1	2	i	$2i$	$1+i$	$1+2i$	$2+i$	$2+2i$
2	0	2	1	$2i$	i	$2+2i$	$2+i$	$1+2i$	$1+i$
i	0	i	$2i$	2	1	$2+i$	$1+i$	$2+2i$	$1+2i$
$2i$	0	$2i$	i	1	2	$1+2i$	$2+2i$	$1+i$	$2+i$
$1+i$	0	$1+i$	$2+2i$	$1+2i$	$2+i$	2	$2i$	i	1
$1+2i$	0	$1+2i$	$2+i$	$2+2i$	$1+i$	i	2	1	$2i$
$2+i$	0	$2+i$	$1+2i$	$1+i$	$2+2i$	$2i$	1	2	i
$2+2i$	0	$2+2i$	$1+i$	$2+i$	$1+2i$	1	i	$2i$	2

The most interesting algebras $A[s]$ are, for our purpose, those in which multiplication is also commutative. They are called *finite fields*.

Finite fields (Galois fields)

Finite fields are algebras $A[s]$ in which the multiplicative group is also Abelian. An algebra $A[s]$ where the order s is a prime is always a field.

The number of integers in a finite field is called its *characteristic*. As in all algebras $A[s]$, this number is a prime, and the order of the field is a power of its characteristic.

Let a be an element of a field of order $p^n = s$, and form its powers $a^0 = 1, a, a^2, \ldots, a^n$. It follows from the proof of Theorem 1.6 that they can be expressed as

$$a^i = c_{i1} w_1 + \ldots + c_{in} w_n,$$

where the w_i are elements of the field, and the c_{ij} are integer elements. Eliminating the w_i from the $n+1$ equations $(i = 0, 1, \ldots, n)$, we obtain $c_0 a^0 + \ldots + c_n a^n = 0$. Thus every element of the field of order p^n satisfies an equation of a degree not larger than n, with integer coefficients.

We show now that in a finite field of order s there is at least one element of order $s-1$ in the multiplicative group, i.e. one whose lowest power equal to 1 is $s-1$. Such an element is called a *primitive element* of the field.

In the multiplicative group every element of the field has an order t such that $x^t = 1$. If follows from Theorem 1.5 that if k is the highest order of any element in the group, then $x^k = 1$ as well. A polynomial of order k cannot have more than k roots (this is proved as in ordinary algebra), and hence we have $s-1 \leqslant k$, because all $s-1$ non-zero elements are roots.

On the other hand, if we multiply all non-zero elements $x_1, \ldots, x_{s-1}$, the result is the same as if we had multiplied $xx_1, \ldots, xx_{s-1}$, where x is a non-zero element. (Note that we use here the commutativity of multiplication.) Hence $x^{s-1} = 1$, and therefore $s - 1 = O(\mathrm{mod}\, k)$. It follows that $s-1 = k$. Thus the highest order of the elements is $s-1$: at least one element has this order, i.e. at least one primitive element exists.

It follows from the definition that the powers of a primitive element up to the $(s-1)$th form all the elements of the field, except 0. Therefore the multiplicative group is cyclic.

Examples

We give here a list of primitive elements of fields of order p:

p	Primitive element	p	Primitive element
3	2	13	2
5	2	17	3
7	3	19	2
11	2	23	5

A primitive element, like any other element of a finite field of order $s = p^n$, satisfies an equation with integer coefficients of degree not larger than n. In fact, a primitive element satisfies an irreducible equation with integer coefficients of degree precisely n, and no equation of lower degree. This will now be shown.

Assume that the irreducible equation of lowest degree satisfied by the primitive element x is

$$c_0 + c_1 x + \ldots + c_k x^k = 0,$$

where c_k is not zero and can, without loss of generality, be assumed to equal 1. Using this equation, any power of x can be expressed as a polynomial of degree not exceeding $k-1$ with integer coefficients which are not all zero. But there are only $p^k - 1$ such polynomials, while there are $p^n - 1$ distinct powers of x; hence $k = n$.

In fact, it can be proved that all roots of this irreducible equation are primitive elements (see, e.g., [15], p. 250), but we do not need this theorem here.

Generally, all powers of an element satisfying an equation

$$h(x) = x^r - (a_0 x^{r-1} + \ldots + a_{r-1}) = 0$$

can be equated to a polynomial of lower degree than r, thus:

$$x^0 = 1, \ldots, x^r = a_0 x^{r-1} + \ldots + a_{r-1}, x^{r+1} = x^r . x, \ldots$$

If $x^a = b_0 x^{r-1} + \ldots + b_{r-1}$, then

$$x^{a+1} = (b_0 a_0 + b_1) x^{r-1} + (b_0 a_1 + b_2) x^{r-2} + \ldots + (b_0 a_{r-2} + b_{r-1})x + b_0 a_{r-1}.$$

(Cf. [90] and [71].)

From Theorem 1.6, the order of a finite field must be the power of a prime. We shall now prove that, conversely, there exists a finite field of order s whenever s is the power of a prime.

The proof uses polynomials in an indeterminate element x. The coefficients in all these polynomials are integers from a field with characteristic p. Let $s = p^n$ and take the polynomial

$$g(x) = x^n + a_1 x^{n-1} + \ldots + a_{n-1}x + a_n.$$

If any polynomial is divided by $g(x)$ and the coefficients of the remainder are reduced mod p, then we say that the resulting remainder is a polynomial which is congruent to the original $g(x)$ modulis $p, g(x)$. There are p^n such polynomials, and they can be obtained by addition and multiplication modd $p, g(x)$.

We ask now whether these polynomials form a field of order p^n. This is certainly not the case if $g(x)$ is reducible, mod p, i.e. if it is the product of two polynomials with integer coefficients, each of degree at least 1. The product $g(x)$ of these two polynomials would be congruent to 0, even though none of the factors is congruent to 0, and this would contradict a property of the algebras.

We assume therefore that $g(x)$ is irreducible, mod p. There is no difficulty in establishing that the rules defining a finite field are valid in this case, except that the existence of an inverse for all polynomials different from 0 deserves some more detailed consideration. It will now be shown that such an inverse exists.

We prove, first, that if two polynomials, $g(x)$ and $f(x)$, say, have no common divisor of degree at least 1, then we can find polynomials $p(x)$ and $q(x)$ such that $f(x) p(x) + g(x) q(x) = 1$.

Consider $f(x) p(x) + g(x) q(x)$ for all polynomials $p(x)$ and $q(x)$, and let $s(x)$ be the polynomial of lowest degree among these.

By division we obtain $h(x)$ such that $f(x) - h(x) s(x) = r(x)$, where $r(x)$ has lower degree than has $s(x)$.

We have then

$$h(x) f(x) p(x) + h(x) g(x) q(x) = f(x) - r(x)$$

for some $p(x)$ and $q(x)$, or also

$$[1 - h(x) p(x)] f(x) + [-h(x) q(x)] g(x) = r(x).$$

But $r(x)$ has lower degree than $s(x)$, and must therefore be 0; otherwise $s(x)$ could not be of smallest degree amongst all $f(x) p(x) + g(x) q(x)$.

It follows that $f(x) = O(\bmod s(x))$. Similarly, $g(x) = O(\bmod s(x))$, so that $s(x)$ is a common divisor of $f(x)$ and $g(x)$.

We have assumed that $f(x)$ and $g(x)$ are relatively prime, and therefore $s(x)$ is a constant, which we can take to be equal to 1.

Let now $g(x)$ be an irreducible polynomial mod p of degree n, and let $f(x)$ be of lower degree. Then there exist polynomials $p(x)$ and $q(x)$ such that $f(x) p(x) + g(x) q(x) = 1$, so that $f(x) p(x) = 1$ (modd $p, g(x)$), i.e. $f(x)$ has an inverse, namely $p(x)$.

We have thus proved that, provided $g(x)$ is irreducible mod p, every polynomial of lower degree has an inverse modd $p, g(x)$. To complete the proof of the existence of a finite field for any prime

power order we must still prove that there exists at least one polynomial, irreducible $\bmod p$, for any degree n. We have to show this,
in fact, only for n exceeding 1, because we know already that fields
with prime order exist. In any case, $x + 1$ is irreducible, whatever
the value of p.

We prove first that if $g(x)$ is an irreducible polynomial $\bmod p$ of
degree $n > 1$ with integer coefficients, then $g(x)$ is a factor of
$x^{s-1} - 1$, $\bmod p$, where $s = p^n$. The residuals $\bmod d$ $p, g(x)$ form a
finite field of order s. The polynomial x is one of the elements, and
hence $x^{s-1} = 1$ $(\bmod d$ $p, g(x))$. It follows that $x^{s-1} - 1$ is equivalent
to zero, and hence to $g(x)$. Therefore $g(x)$ is a factor of $x^{s-1} - 1$.

Let $f(x) = a_0 + a_1 x + \ldots + a_k x^k$ be a polynomial with integer coefficients $\bmod p$. Since

$$\frac{p!}{b_1! \ldots b_k!} \quad \text{where } b_1 + \ldots + b_k = p, \text{ and all } b_i \geqslant 0,$$

equals 1 if $p = b_i$ for some i, and equals $O \pmod{p}$ otherwise,
we have

$$[f(x)]^p = a_0^p + a_1^p x^p + \ldots + a_k^p x^{kp} = a_0 + a_1 x^p + \ldots + a_k x^{kp} = f(x^p)$$

(because $a_i^p = a_i$, $\bmod p$).

By induction, $[f(x)]^s = f(x^s)$, where $s = p^n$.

We can now prove that an irreducible polynomial $g(x)$ of degree
$t > n$ cannot be a divisor of $x^{s-1} - 1$, where $s = p^n$. The residuals
$\bmod d$ $p, g(x)$ form a finite field of order p^t. Let the primitive element
be $f(x) = a_0 + a_1 x + \ldots + a_k x^k$, $k < t$. If $g(x)$ were a divisor of $x^{s-1} - 1$,
then $[f(x)]^s = f(x^s) = f(x)$, $\bmod d$ $p, g(x)$, and we could not obtain
from the powers of $f(x)$ more than $s - 2$ different elements, because
the $(s-1)$th power would be unity. But the field based on $g(x)$ has
$p^t - 1$ non-zero elements, and since $p^t > p^n = s$, we have reached a
contradiction.

Imagine now $x^{s-1} - 1$ decomposed into its irreducible factors.
None of these has a degree larger than n, as we have just seen. Also,
a polynomial $x^m - 1$ has no double root $\bmod p$, unless $m = O(\bmod p)$.
(This is shown as in differential calculus, using the fact that $x^m - 1$
and $m x^{m-1}$ have no common factor $\bmod p$ if m is relatively prime to
p.) All irreducible polynomials of degree $f < n$ are factors of $x^{p^f - 1} - 1$,
and since $x^{p^n - 1} - 1$ has no double roots, the sum of the degrees of the
irreducible factors of degree f is at most $p^f - 1$.

Hence the sum of the degrees of all the factors of degree less than n is smaller than $p + p^2 + \dots + p^{n-1}$, which in its turn is smaller than $(p^n-1)/(p-1)$. Thus there must also exist at least one irreducible factor of degree n.

The proof that there exists a finite field of order s for any prime power s is thus complete.

A finite field with polynomials modd p, $g(x)$ as elements is called a *Galois field*. If its order is s, then we denote it by $GF(s)$.

Examples

(1) Consider the $GF(3^2)$. To construct it, we must find an irreducible polynomial $g(x)$ with coefficients mod 3 and of degree 2. We have $x^8-1 = (x+1)(x+2)(x^2+1)(x^2+x+2)(x^2+2x+2)$. Any one of the last three factors can serve as a $g(x)$. We use the first of these, and write the elements of the resulting $GF(9)$ using i for x, as a reminder that because $x^2+1 = 0$, addition and multiplication reflect those of the square root of -1, mod 3. The elements are:

$$0 \quad 1 \quad 2 \quad i \quad i+1 \quad i+2 \quad 2i \quad 2i+1 \quad 2i+2.$$

From $x^2+1 = 0$, all powers of i can be equated to a polynomial in i of degree less than 2, as follows:

i^0	i^1	i^2	i^3	i^4	i^5	i^6	i^7
1	i	2	$2i$	1	i	2	$2i.$

We note that i is not a primitive element. However, the roots of $x^2+x+2 = 0$, namely $i+1$ and $2i+1$, and also those of $x^2+2x+2 = 0$, namely, $i+2$ and $2i+2$, are primitive elements. As an illustration, we equate the powers of $x = i+1$ to polynomials of degree less than 2:

x^0	x^1	x^2	x^3	x^4	x^5	x^6	x^7	x^8
1	x	$2x+1$	$2x+2$	2	$2x$	$x+2$	$x+1$	1.

(2) Consider a $GF(2^4)$. An irreducible polynomial is, for instance, x^4+x^3+1. The 15 non-zero elements of $GF(2^4)$ correspond to the root x of this polynomial as follows:

x^0	x^1	x^2	x^3	x^4	x^5	x^6	x^7
1	x	x^2	x^3	x^3+1	x^3+x+1	x^3+x^2+x+1	x^2+x+1

x^8	x^9	x^{10}	x^{11}	x^{12}	x^{13}	x^{14}	x^{15}
x^3+x^2+x	x^2+1	x^3+x	x^3+x^2+1	$x+1$	x^2+x	x^3+x^2	1

Theorem 1.7 Any two fields of the same finite order are isomorphic.

Hence, in an abstract sense, there is only one field of a given finite order. We do not lose anything by talking about a Galois field $GF(s)$, whenever we mean a finite field of order s.

Proof: We show that a finite field F of order s is isomorphic to a given $GF(s)$. Let the field to be studied have the elements $x_0 = 0$, $x_1, ..., x_{s-1}$. All these except x_0 satisfy the equation $x^{s-1} - 1 = O(\bmod p)$, so that we have, $\bmod p$,

$$x^{s-1} - 1 = (x - x_1) \ldots (x - x_{s-1}).$$

Let the given $GF(s)$ be constructed from the $\bmod p$ irreducible polynomial $g(x)$ of degree n. The latter is a divisor of $x^{s-1} - 1$, so that there must be some element, say x_i, such that $g(x_i) = 0$.

Let the elements of the $GF(s)$ be polynomials $b_0 x^{n-1} + ... + b_{n-1}$. The polynomials in x_i, $b_0 x_i^{n-1} + ... + b_{n-1}$, where the coefficients are integers in F, are all different, because if two of them were equal, then their difference, of lower degree than n, would have a factor $(x - x_i)$ in common with $g(x)$, and the latter would not be irreducible. Thus these polynomials form all the elements of F. The correspondence of $b_0 x_1^{n-1} + ... + b_{n-1}$, an element of F, and of $b_0 x^{n-1} + ... + b_{n-1}$ ($\bmod\bmod p, g(x)$), an element of $GF(s)$, is clearly an isomorphism. This completes the proof.

Example

Let F contain the elements $0, 1, 2, i, i+1, i+2, 2i, 2i+1, 2i+2$, whose addition and multiplication are carried out, $\bmod 3$, as if i were the square root of -1. Let, moreover, $g(x) = x^2 + x + 2$. The element $i+1$ is a root of $g(x) = 0$. This establishes the isomorphism

0	1	2	i	$i+1$	$i+2$	$2i$	$2i+1$	$2i+2$
0	1	2	$1+x$	$2+x$	x	$2+2x$	$2x$	$1+2x$.

Lists of irreducible polynomials can be found, for instance, in [13] and in [14], quoted in [1] and in [15]. We mention here a few:

$$p = 2 \quad n = 2 \quad x^2 + x + 1$$
$$3 \quad x^3 + x^2 + 1, \; x^3 + x + 1$$
$$4 \quad x^4 + x^3 + 1, \; x^4 + x + 1$$
$$p = 3 \quad n = 2 \quad x^2 + 1, \; x^2 + x + 2, \; x^2 + 2x + 2$$
$$3 \quad x^3 + 2x + 1$$
$$4 \quad x^4 + x^3 + x^2 + 2x + 2.$$

A method for constructing such polynomials by automatic computation is described in [99].

In our proof of the existence of a Galois field of degree $s = p^n$ we have used polynomials of degree n with coefficients which were elements of a $GF(p)$. Now that we know that fields of any prime power order exist, we can repeat the whole argument using irreducible polynomials $g(x)$ of degree m, with coefficients from a $GF(s)$, where $s = p^n$; we then find that a $GF(s^m)$ is a set of elements represented by polynomials of degree less than m, their coefficients being elements from a $GF(s)$. We shall then conclude that a primitive element x of the $GF(s^m)$ satisfies an irreducible equation of degree $m-1$ with coefficients from the $GF(s)$. All powers of x can be represented as polynomials in x of a degree less than m. The 0-th, 1st, ..., $(s^m - 2)$th powers are all different, and

$$x^{s^m-1} = 1.$$

The powers $x^{jt} (j = 0, 1, ..., s-2)$ of x, where $t = (s^m - 1)/(s-1)$, form a $GF(s)$. Thus a $GF(s^m)$ contains a $GF(s)$. Conversely, a $GF(p^n)$ contains a $GF(p^k)$ only if k is a factor of n, because a primitive element of $GF(p^k)$ has order $p^k - 1$, and this must be a factor of $p^m - 1$. This is so if and only if k is a factor of m.

For $m = 1$ we obtain our earlier results.

In the following paragraphs we consider some further properties of Galois fields, which will be of use in later chapters of this volume.

Let x be a primitive element of $GF(s)$ and consider the powers $x^0, x^1, ..., x^{s-2}$. Any even power is clearly the square of some element. But does there exist an element y whose square is equal to an odd power of the primitive element? For any y there exists a u, not exceeding $s-2$, such that $y = x^u$, $y^2 = x^{2u}$, and if $2u > s-2$, then we write $y^2 = x^{2u-v(s-1)}$, where v is an integer such that $0 \leqslant 2u - v(s-1) \leqslant s-2$.

If s is odd, then $2u - v(s-1)$ is even, so that the odd powers of the primitive element are not squares. On the other hand, if s is even, i.e. a power of 2, then every non-zero element is a square, because then

$$x^{2u} = (x^u)^2 \quad \text{for } u = 0, 1, ..., (s-2)/2, \quad \text{and}$$
$$x^{2u+1} = (x^{u+s/2})^2 \quad \text{for } u = 0, 1, ..., (s-2)/2.$$

This accounts for all powers of x.

We have thus found that, when s is odd, there are in $GF(s)$ $(s-1)/2$ non-zero squares, and the same number of non-squares.

Example

In $GF(9)$, constructed from $x^2 + 1$, the non-zero squares are $1, 2,$ i, and $2i$, and the other non-zero elements are not squares.

The product of two squares, and that of two non-squares, is a square. That of a non-zero square and a non-square is not a square. This follows from the fact that the sum of two even numbers, and that of two odd numbers, is even, while that of an even number and of an odd number is odd. (Write the elements as powers of a primitive element!)

We use the "Lagrange symbol" χ, defined as follows: $\chi(a) = 1$ if a is a non-zero square, $\chi(a) = -1$ if it is a non-square, and $\chi(0) = 0$. Then what we have just proved can be stated as follows:

If s is odd, then

(I) $\sum \chi(a) = 0$ (summation over all elements of $GF(s)$).

(II) $\chi(a) . \chi(b) = \chi(a . b)$.

All powers of a primitive element x are distinct up to the $(s-1)$th, and $x^{s-1} = 1$. Therefore, when s is odd, $x^{(s-1)/2} = -1$. Hence -1 is a square when $(s-1)/2$ is even, and a non-square otherwise. We have thus, for odd s,

(III) $\chi(-1) = 1$ when $s = 4t+1$, and $= -1$ when $s = 4t+3$, where t is an integer.

We shall also need the following theorem:

(IV) When s is odd, then $\sum_j \chi(j - i_1) . \chi(j - i_2) = -1$, when $i_1 \neq i_2$. The summation extends over all elements of $GF(s)$.

Proof (cf. [67]): By (II), the left-hand side equals

$$\sum_j \chi((j - i_1)(j - i_2)).$$

Introduce $u_0 = (i_1 - i_2)/2 \leqslant 0$, and $v = [j - (i_1 + i_2)/2]/u_0$. The left-hand side becomes, after simple algebraic transformations,

$$\sum_v \chi(u_0^2(v^2 - 1))$$

where v is summed over all elements of $GF(s)$. This is, again by (II),

$$\sum_v \chi(u_0^2) . \chi(v^2 - 1) = \sum_v \chi(v^2 - 1).$$

We count how many of the terms in the last sum are positive, negative, or zero. For the symbol to have value $+1$, we must have some x such that $v^2 - 1 = x^2$, i.e. $(v+x)(v-x) = 1$, $v - x = (v+x)^{-1}$. Let $v + x = y$, then $v = (y + y^{-1})/2$, $x = (y - y^{-1})/2$. Therefore we have

$X(v^2-1) = 1$ as often as v can be expressed in the form $(y+y^{-1})/2$. (Note that if $y+y^{-1} = w+w^{-1}$, then $(y-w)(1-yw) = 0$, so that either $y = w$, or $y = w^{-1}$.)

If $y = 1$, or $= -1$, then $X(v^2-1) = 0$. Apart from these two cases y and y^{-1} are distinct, so that $X(v^2-1) = 1$ as often as there are distinct pairs y, y^{-1}, ignoring 1 and -1. There are $(s-1)/2 - 1 = (s-3)/2$ such cases, and there remain $(s-1)/2$ values of v for which $X(v^2-1) = -1$. Adding all terms together and cancelling 1 against -1, there remains a single -1, which proves (IV).

Hadamard matrices

We investigate now a special matrix, which we call a *Hadamard matrix* (see [24], and also [64]). Such a matrix H_n is defined as an orthogonal matrix of order n, whose elements are 1 and -1. For instance,

$$H_2 = \begin{pmatrix} 1 & 1 \\ 1 & -1 \end{pmatrix}.$$

We have $H_n H_n' = nI_n$, and hence $H_n^{-1} = H_n'/n$, and $H_n H_n' = H_n' H_n$. Also, the determinant $|H_n H_n'| = n^n$, so that $|H_n| = n^{n/2}$.

Such matrices are of interest in various branches of mathematics. For instance, the absolute value of a real determinant with elements in the range -1 to 1 cannot be larger than $n^{n/2}$, and the latter value is only attained for a Hadamard matrix of order n.

It is easily seen that the order of a Hadamard matrix (a_{ij}) can only be 1, 2, or a multiple of 4. We have

$$\sum_{j=1}^{n} (a_{1j} + a_{2j})(a_{1j} + a_{3j}) = \sum_{j=1}^{n} a_{1j}^2 = n.$$

Every term in the first sum is either 0 or 4, and the result follows. It has been conjectured that there exist H_n for all n divisible by 4, and all those up to 112 have been constructed (see [81]). All those up to $n = 100$, excluding 92, are given implicitly in [67], and that for 92 in [4]. A list of Hadamard matrices known to exist is contained in [97].

Let $s = p^n$ be of the form $4t-1$, where p is a prime and n and t are positive integers. The following matrix is then of Hadamard type:

$$\begin{pmatrix} 1 & 1 & 1 & . & . & . & 1 & 1 \\ 1 & -1 & X(1) & . & . & . & X(s-2) & X(s-1) \\ 1 & X(s-1) & -1 & . & . & . & X(s-3) & X(s-2) \\ 1 & X(1) & X(2) & . & . & . & X(s-1) & -1 \end{pmatrix}$$

where the arguments $1, 2, \ldots, s-1$, stand for the elements of $GF(s)$.

The inner product of the first and the $(i+1)$th rows is

$$1 - 1 + \sum_{j=1}^{s} X(j-i) = 0, \quad \text{by (I)}.$$

The inner product of the (i_1+1)th and the (i_2+1)th rows, where $i_1 \neq i_2$, is

$$1 - X(i_1-i_2) - X(i_2-i_1) + \sum_{j=1}^{s} X(j-i_1) X(j-i_2)$$
$$= -X(i_1-i_2) - X(i_2-i_1) \quad \text{by (IV)}.$$

But $X(i_1-i_2) = X(-1) . X(i_2-i_1)$, and $X(-1) = -1$ by (III), so that the inner product of the two rows is again 0.

If H_n is a Hadamard matrix of order n, then $\begin{pmatrix} H_n & H_n \\ H_n & -H_n \end{pmatrix}$ is one of order $2n$. Therefore all matrices of order $2^h(p^n+1)$ can be obtained in this manner, where p^n is of the form $4t-1$, or may be replaced by 0.

If $s = p^n$ is of the form $4t+1$, then a Hadamard matrix of order $2(s+1) = 4(2t+1)$ can be obtained as follows: consider the matrix of order $s+1$

$$X = (b_{ij}) = \begin{pmatrix} 0 & 1 & 1 & . & . & . & 1 & 1 \\ 1 & 0 & X(1) & . & . & . & X(s-2) & X(s-1) \\ 1 & X(s-1) & 0 & . & . & . & X(s-3) & X(s-2) \\ 1 & X(1) & X(s-1) & . & . & . & X(s-1) & 0 \end{pmatrix}.$$

This is, of course, not a Hadamard matrix, because the value 0 appears; however, it is orthogonal, because

$$\sum_{j=1}^{s} X(j - i_k) = 0, \quad \text{by (I), and}$$

$$1 + \sum_{j=1}^{s} X(j - i_1) X(j - i_2) = 0, \quad \text{by (IV)}.$$

If we replace 1 by $C = \begin{pmatrix} 1 & 1 \\ 1 & -1 \end{pmatrix}$, -1 by $-C = \begin{pmatrix} -1 & -1 \\ -1 & 1 \end{pmatrix}$, and 0 by

$D = \begin{pmatrix} 1 & -1 \\ -1 & -1 \end{pmatrix}$, then we obtain a matrix of order $2(s+1)$ which is of Hadamard type. We see this as follows:

The inner product of two rows originating from the same row of X is zero, by their definition. We have to show that the inner product of any other pair of rows is also zero. This can be expressed by saying that $M_1 M_2' = \begin{pmatrix} 0 & 0 \\ 0 & 0 \end{pmatrix}$, where M_1 and M_2 are two matrices arising from the i_1-th and the i_2-th row of X respectively.

If these rows of X are

$$(b_{i_1 0}, \ldots, b_{i_1 s}) \quad \text{and} \quad (b_{i_2 0}, \ldots, b_{i_2 s}),$$

then M_1 is $(b_{i_1 0}C, \ldots, b_{i_1 s}C)$, where however a term 0 is replaced by D, and similarly for M_2. Therefore

$$M_1 M_2' = \sum_{j=1}^{s}(b_{i_1 j}C)(b_{i_2 j}C') + D(b_{i_2 i_1}C) + (b_{i_1 i_2}C)D',$$

$j \neq i_1, i_2$. Because s is of the form $4t+1$, this equals

$$CC' \sum_{j=1}^{s} b_{i_1 j}b_{i_2 j} + X(i_1 - i_2)(DC' + CD') = CC'.0 + X(i_1 - i_2)\begin{pmatrix} 0 & 0 \\ 0 & 0 \end{pmatrix}$$

$$= \begin{pmatrix} 0 & 0 \\ 0 & 0 \end{pmatrix}, \quad \text{using (IV)}.$$

We have thus proved that Hadamard matrices exist for all orders $2^h(p^n + 1)$, where p is either zero or any odd prime.

Difference sets

A set of integers $d_1, \ldots, d_k$ is said to be a difference set $\bmod v$, if the $k(k-1)$ differences $d_i - d_j$ equal any non-zero value $\bmod v$ the same number (say λ) of times. It follows at once that $k(k-1) = \lambda(v-1)$.

The $2\lambda + 1$ non-zero quadratic residues of a prime number (i.e. remainders of squares after division by the prime number) of the form $v = 4\lambda + 3$ form such a difference set $\bmod v$, as we shall now show.

Let $a^2 - b^2 = 1$, $ab \neq 0$, have λ solutions. Then $A^2 - B^2 = d$, $AB \neq 0$, where $d \neq 0$, has the same number of solutions, because if d is a square, say $d = u^2$, then $A = au$, $B = bu$ is a solution, and if d is not a square, then $-d$ is a square (by (III), page 18), and hence $A = bu$, $B = au$ is a solution.

Consequently, all values $x^2 - y^2$ (where x and y are residues mod v, and $xy \neq 0$, $x \neq y$, produce all non-zero differences equally often. The number of differences $x^2 - y^2$ is

$$2 \binom{(v-1)/2}{2} = (v-1)(v-3)/4.$$

There are $v - 1$ non-zero differences, so that the number of repetitions is $\lambda = (v - 3)/4$.

Examples

(1) $v = 7$, primitive element 3. A difference set is $3^2 = 2$, $3^4 = 4$, $3^6 = 1$, and $\lambda = 1$.

(2) $v = 11$, primitive element 2. A difference set is $1, 3, 4, 5, 9$, and $\lambda = 2$.

S. Chowla has shown in [16] that quartic residues mod $4x^2+1$ (prime), where x is odd, form difference sets, and this, with other similar constructions, is also proved in [49].

Difference sets can also be found from the properties of finite spaces (see Chapter III).

Evans and Mann have shown in [21] that for $k \leqslant 1600$, difference sets with $\lambda = 1$ only exist if $k - 1$ is a prime power.

A comprehensive survey of difference sets is contained in [29]. The smallest value of k for which non-isomorphic solutions exist for the same v is $k = 15$. It is shown that there are at least two non-isomorphic solutions for $v = p$, $k = (p-1)/2$, $\lambda = (p-3)/4$, whenever p is a prime of the form $4x^2 + 27$, and four non-isomorphic solutions are known for $v = 121$, $k = 40$, $\lambda = 13$. A more general fact is proved in [23]: let q be a prime power, n and m positive integers where n is at least 3, and m is the product of r (not necessarily distinct) prime numbers. Then there are at least 2^r non-isomorphic perfect difference sets with

$$v = (q^{nm} - 1)/(q - 1), \quad k = (q^{nm-1} - 1)/(q - 1),$$

$$\lambda = (q^{nm-2} - 1)/(q - 1).$$

Methods for combining these sets into other difference sets are also given.

We speak of a system of difference sets if the sets

$$d_{11}, \ldots, d_{1k}$$
$$- \quad - \quad -$$
$$d_{m1}, \ldots, d_{mk}$$

where the d_{ij} are elements of a group of order s, are such that if we compute all the $mk(k-1)$ values $d_{ij_1}^{-1} d_{ij_2}$ $(i = 1, \ldots, m; j_1 \neq j_2)$ of any two elements in the same set, then each of the $s-1$ non-unit elements appears the same number (say λ) of times. Multiplication is again to be interpreted in the sense as used in the group. We have, of course, $mk(k-1) = \lambda(s-1)$.

Sprott has constructed such systems in [93] for $s = 2t(2\lambda + 1) + 1$, provided this is a prime power. It consists of the sets

$$x^i, \ x^{i+2t}, \ x^{i+4t}, \ \ldots, \ x^{i+4\lambda t},$$

where $i = 0, 1, \ldots, t - 1$, and x is a primitive element of $GF(s)$.

If $\lambda = 1$, then $s = 6t+1$, and each set consists of 3 elements.

[93] contains other constructions as well, and so does [95]. In [87] it is shown that if $n = 4t+3$, then a difference set mod $v = nm$, such that among the $k(k-1)$ differences all integers congruent to multiples of n occur λ_1 times, and all others λ_2 times, can only exist if $-n$ is a quadratic residue of any prime factor contained to an odd degree in $k + (m-1)\lambda_1 - m\lambda_2$. For $\lambda_1 = \lambda_2$ this was already proved in [17].

If a difference set $d_1, \ldots, d_k \bmod v$ is given and there exists a value t such that $td_1, \ldots, td_k$ is equal to $d_1 + s, \ldots, d_k + s$ for some s, except that the elements within the set might have been permuted, then t is called a *multiplier* of the set. Such a multiplier must be relatively prime to v, because for some d_i, d_j we must have $td_i - td_j = 1 \pmod{v}$. The product of two multipliers is, of course, also a multiplier.

Example

2 is a multiplier of $0, 1, 6, 8, 18 \pmod{21}$.

We shall not deal here with the extensive literature on multipliers, except for quoting, without proof, the following theorem (see [26], [33], [63]).

If a prime p, larger than λ, divides $k - \lambda$ but not v, then p is a multiplier of the difference set with parameters v, k, λ.

The condition $p > \lambda$ is possibly superfluous, but no proof of this is known as yet, except in certain special cases mentioned in [63].

We introduce the notation $b_i = d_{i+1} - d_i \pmod{v}$ $(i = 1, \ldots, k)$ where the subscripts are to be taken $\mod k$. The "circular sums" $b_s + b_{s+1} + \ldots + b_{t-1}$ are then congruent, $\mod v$, to $d_t - d_s$. Because there are λ differences equal to any given value, this is also true of the $k(k-1)$ circular sums. When $\lambda = 1$, then the number of these sums equals $v - 1$. If we count also the sum $b_1 + \ldots + b_k$ (which is $O(\mod v)$) then we obtain (in this case, when $\lambda = 1$)

$$v = 1 - k + k^2 = 1 + (k-1) + (k-1)^2$$

circular sums of $b_1, \ldots, b_k \,(\mod v)$. The b_i are called a *perfect partition* of v (see [48]).

Example

$$v = 13 \qquad d_i \quad 0 \ 1 \ 3 \ 9$$
$$b_i \quad 1 \ 2 \ 6 \ 4 \ \text{(adding up to } 13 = v)$$

circular sums:

$1 = 1$	$2 = 2$	$6 = 6$	$4 = 4$
$+2 = 3$	$+6 = 8$	$+4 = 10$	$+1 = 5$
$+6 = 9$	$+4 = 12$	$+1 = 11$	$+2 = 7$
$+4 = 0$	$+1 = 0$	$+2 = 0$	$+6 = 0.$

EXERCISES

(1) Find the regular permutation group which is isomorphic to the group consisting of the following matrices under multiplication:

$$\begin{pmatrix} 1 & 0 \\ 0 & 1 \end{pmatrix} \quad \begin{pmatrix} -1 & 1 \\ -1 & 0 \end{pmatrix} \quad \begin{pmatrix} 0 & -1 \\ 1 & -1 \end{pmatrix} \quad \begin{pmatrix} 0 & 1 \\ 1 & 0 \end{pmatrix} \quad \begin{pmatrix} -1 & 0 \\ -1 & 1 \end{pmatrix} \quad \begin{pmatrix} 1 & -1 \\ 0 & -1 \end{pmatrix}.$$

(2) Find the regular permutation group which is isomorphic to the group consisting of the following matrices under multiplication:

$$\begin{pmatrix} 1 & 0 \\ 0 & 1 \end{pmatrix} \quad \begin{pmatrix} -1 & 1 \\ -1 & 0 \end{pmatrix} \quad \begin{pmatrix} 0 & -1 \\ 1 & -1 \end{pmatrix} \quad \begin{pmatrix} -1 & 0 \\ 0 & -1 \end{pmatrix} \quad \begin{pmatrix} 1 & -1 \\ 1 & 0 \end{pmatrix} \quad \begin{pmatrix} 0 & 1 \\ -1 & 1 \end{pmatrix}.$$

(3) Construct a $GF(4)$ and its tables of multiplication and of addition.

(4) Construct a difference set modulo 19, $\lambda = 4$.

CHAPTER II

FINITE PLANES

Definition

A finite projective plane is defined as a finite set of elements, called *points*, and subsets of the points, called *lines*, subject to the following conditions:-

(1) To every pair of points there is exactly one line which contains both points.

(2) To every pair of lines there is one point contained in both lines, called their *intersection*.

From the latter condition, and (1), it follows that there is exactly one such point. If there were two, then both lines would contain them, and this would contradict (1).

(3) There are in the set four points such that no three of them are on the same line.

It follows that each line contains at least three points, and that through each point there must be at least three lines. For instance, if the four points mentioned are P,Q,R,S, then the line through P and Q will also contain its intersection with the line through R and S, and a similar proof holds for other lines, and for the second part of the statement.

Consider a line **a** with $s + 1$ points on it; let one of them be A. Consider the further lines **b** and **c** through A. On **c**, there will be a point different from A, say C. Connect C and the s points on **a** which are different from A. These lines will intersect **b** in s distinct points. There cannot be more than $s + 1$ points on **b**, because if there were, then such a further point, connected to C, would intersect **a** in a further point. Thus there will be precisely $s + 1$ points on any line. There will also be precisely $s + 1$ lines through C and, by an argument similar to that just used, precisely $s + 1$ lines through any point.

There are $s + 1$ lines through A, and they contain between them all the points of the plane. Apart from A, there are s on all of them, hence altogether $1 + s(s + 1) = s^2 + s + 1$ in the plane, and similarly,

s^2+s+1 lines. The latter fact follows most easily from the consideration that in all defining conditions (axioms) above, and hence also in
all their consequences, concepts of line and point can be exchanged.

Example

The simplest finite projective plane is that with $s = 2$; there are
precisely three lines through each point, and three points on each line.
Altogether, there are 7 points and 7 lines in the plane. It may be illustrated by Fig. 1.

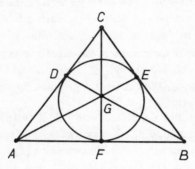

Fig. 1

The points are A, B, C, D, E, F and G and the lines are ADC, AGE,
AFB, CGF, CEB, DGB and DEF. The last-named three points are not
connected by a straight line, but straightness is not a meaningful concept in a finite plane. Here a line is merely a subset of points.

Construction from algebras

We shall now study ways in which finite projective planes can be
constructed from algebras $A[s]$.

Define a point as an ordered set of three "homogeneous co-ordinates"
(x_0, x_1, x_2), where the x_i are elements of $A[s]$, not all of them simultaneously 0. The set of co-ordinates (ax_0, ax_1, ax_2) where a is also
an element of $A[s]$, different from 0, shall denote the same point. Any
other two triplets define two different points. Thus each point is
represented in $s - 1$ ways, and because there are $s^3 - 1$ possible triplets of co-ordinates, the total number of different points is $(s^3-1)/(s-1)$
$= s^2+s+1$.

We define a line as the set of all those points whose co-ordinates
satisfy an equation of the form $x_0 + x_1 a_1 + x_2 a_2 = 0$, or of the form

$x_1 + x_2 a_2 = 0$, or of the form $x_2 = 0$, where the a_j are also elements of $A[s]$. It follows from the distribution law that if (x_0, x_1, x_2) satisfies such an equation, then (ax_0, ax_1, ax_2) satisfies the same equation. There are $s^2 + s + 1$ such lines in the plane, and we have to show that the points and lines thus defined satisfy the axioms of a finite projective plane.

For the proof, we take standard representations of points, as follows:– a point with $x_2 \neq 0$ is represented by $(x_0, x_1, 1)$, a point with $x_2 = 0$ and $x_1 \neq 0$ by $(x_0, 1, 0)$, and a point with $x_1 = x_2 = 0$ by $(1, 0, 0)$.

(1)   *Through any two points there is one line.*

We give the proof for points of the first two types. Let the points have co-ordinates $(x_0', x_1', 1)$ and $(y_0', 1, 0)$. Then the line through these two points has the equation –

$$x_0 + x_1 (-y_0') + x_2(-x_0' + x_1' (-y_0')) = 0$$

In the same way, we can find the connecting line for any other two points.

(2)   *Any pair of lines meet in one point.*

The proof is analogous to that of (1).

(3)   Of the following four points, it is easily seen that no three are on the same line:

$$(1, 0, 0) \quad (0, 1, 0) \quad (0, 0, 1) \quad (1, 1, 1).$$

If, in particular, a projective plane is derived from a $GF(s)$, then it has properties which do not necessarily apply to one derived from a general $A[s]$.

In that case the equation of any line can be written $x_0 a_0 + x_1 a_1 + x_2 a_2 = 0$ or $a_0 x_0 + a_1 x_1 + a_2 x_2 = 0$, where the a_i are not simultaneously 0, and $(aa_0) x_0 + (aa_1) x_1 + (aa_2) x_2 = 0$, is, of course, the same line. The line connecting the points (y_0, y_1, y_2) and (z_0, z_1, z_2) may then also be defined as the set of all points with co-ordinates $(ay_0 + bz_0, ay_1 + bz_1, ay_2 + bz_2)$, where a and b are elements of $GF(s)$, not both equal to 0. There are $s^2 - 1$ such pairs, and since simultaneous multiplication of a and b by the same non-zero element produces the same set of points, $s + 1$ of them will be different.

If (y_0, y_1, y_2) and (z_0, z_1, z_2) are solutions of $a_0 x_0 + a_1 x_1 + a_2 x_2 = 0$ then a point whose co-ordinates are linear combinations of the co-ordinates of these two points will satisfy the same equation (provided commutativity of multiplication holds); on the other hand, if $x_0 = ay_0 + bz_0$, $x_1 = ay_1 + bz_1$, $x_2 = ay_2 + bz_2$, then (x_0, x_1, x_2) is a point on the line with equation

$$(y_1 z_2 - y_2 z_1)x_0 + (y_2 z_0 - y_0 z_2)x_1 + (y_0 z_1 - y_1 z_0)x_2 = 0$$

whatever the values of a and b. Hence the two definitions of a line are equivalent.

This equivalence does not hold in a plane derived from an algebra where multiplication is not commutative. For instance, if we take the $A[9]$ given in Chapter I, then the line $x_0 + x_1(1+i) + x_2(2+i) = 0$ contains the points P_1 $(1+i, 2, 0)$ and $P_2(2, 1+2i, 2+2i)$, but it does not contain either $(1+i)P_1 + (1+2i)P_2 = (2, 1+2i, 2i)$ or $P_1(1+i) + P_2(1+2i) = (2, 1+2i, i)$. This can be verified from the multiplication table given in Chapter I.

A finite projective plane derived from a $GF(s)$ is denoted by $PG(2, s)$ ("projective geometry"). The plane illustrated in Fig. 1 is a $PG(2, 2)$. The order s of the $GF(s)$ is the order of the $PG(2, s)$, and 2 is called its *dimension*.

Example

$PG(2, 3)$ with 13 points and 13 lines.

The points on such a plane are —

$A(100)$	$a(211)$	$Q(120)$	$X(011)$	$P(111)$
$B(010)$	$b(121)$	$R(102)$	$Y(101)$	
$C(001)$	$c(112)$	$S(012)$	$Z(110)$	

and the lines, with the points they contain, are —

$x_0 = 0$ $BCSX$,	$x_0 + x_1 = 0$ $CabQ$,	$x_0 + x_1 + 2x_2 = 0$ $cQXY$
$x_1 = 0$ $ACRY$,	$x_0 + x_2 = 0$ $BacR$,	$x_0 + 2x_1 + x_2 = 0$ $bRXZ$
$x_2 = 0$ $ABQZ$,	$x_1 + x_2 = 0$ $AbcS$,	$2x_0 + x_1 + x_2 = 0$ $aSYZ$
$x_1 + 2x_2 = 0$ $AaPX$,	$x_0 + 2x_2 = 0$ $BbPY$,	$x_0 + 2x_1 = 0$ $CcPZ$
	$x_0 + x_1 + x_2 = 0$ $PQRS$	

To illustrate this plane, draw a diagram (Fig. 2).

Triangle abc Within this triangle, choose a point P. Draw the lines Pa, Pb, Pc and their intersections with bc (i.e. A), ac (i.e. B) and ab (i.e. C), respectively. Draw the lines AB, AC, BC and their intersections with ab (i.e. Q), ac (i.e. R) and bc (i.e. S), respectively.

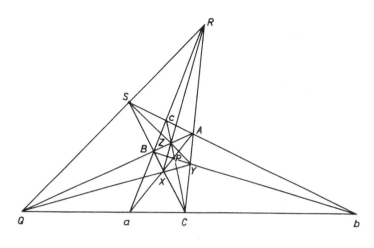

Fig. 2

By the theorem of Desargues (see below, but here applied in ordinary projective geometry) Q, R, and S lie on a straight line. Define X, Y, and Z respectively as the intersections of AP (or Aa) and BC, BP (or Bb) and AC, CP (or Cc) and AB. Then the triples of points YXQ, YZS, XZR will each lie on a straight line. This follows again from the theorem of Desargues. We show it here for X, Y and Q. The proofs for the other triples are analogous.

Consider the triangles ABC and baP.

The lines Ab, Ba and CP all pass through c. Hence the intersections of AB and ba (i.e. Q), AC and bP (i.e. Y) and BC and aP (i.e. X) lie on a straight line.·

The diagram contains 13 points and 13 lines. To make it an illustration of the $PG(2,3)$ we must consider P as lying on QRS, a on SXZ, b on RXZ and c on QXY. Then every line contains 4 points and every point lies on 4 lines.

Diagonal points

We exhibit now a remarkable distinction between the properties of a $PG(2, 2^n)$ and a $PG(2, p^m)$ where $p \neq 2$. The diagonal points of a complete quadrangle are collinear if and only if the order of the finite projective plane is even.

Proof: We may assume, without loss of generality, that the vertices of the quadrangle are $(1, 0, 0)$, $(0, 1, 0)$, $(0, 0, 1)$ and $(1, 1, 1)$. Its six sides are $x_2 = 0$, $x_1 = 0$, $x_1 - x_2 = 0$, $x_0 = 0$, $x_0 - x_2 = 0$, and $x_0 - x_1 = 0$, while the three diagonal points are $(1, 1, 0)$, $(1, 0, 1)$ and $(0, 1, 1)$. The line through the first two points contains all points with co-ordinates $(a + b, a, b)$, and the third point is one of these if and only if $a = b$ and $a + b = 0$. In a $GF(p^m)$ this is only possible if the characteristic of the Galois field is 2.

The latter case is illustrated in Fig. 1. Let the vertices have the co-ordinates $C(1, 1, 1)$, $D(1, 0, 0)$, $E(0, 0, 1)$, $G(0, 1, 0)$ and the diagonal points $A(0, 1, 1)$, $B(1, 1, 0)$, $F(1, 0, 1)$. In this case the diagonals are the lines CGF, AFB and DEF.

A plane in which all complete quadrangles have collinear diagonal points is called a *Fano-plane*.

The theorem of Desargues

In a $PG(2, s)$ the following theorem, named after Girard Desargues, holds (since the theorem is valid in ordinary projective geometry, it can be illustrated by drawing straight lines; see Fig. 3).

If two triangles, not degenerating into straight lines, and without common vertices, say ABC and abc, are such that the lines Aa, Bb and Cc pass through the same point, say P, then and only then the intersections of the lines AB and ab, of AC and ac, and of BC and bc are collinear.

Proof: Let, without loss of generality, the vertices of the first triangle be $A(1, 0, 0)$, $B(0, 1, 0)$, $C(0, 0, 1)$ and let P have the co-ordinates (f, g, h). Then the co-ordinates of a, b, c, are respectively of the form $(f + u, g, h)$, $(f, g + v, h)$ and $(f, g, h + w)$. The actual values of u, v, and w depend on the position of a, b, and c, but none of them equals 0.

The line through bc has the equation $(vh + gw + vw)x_0 - fwx_1 - fvx_2 = 0$, and it meets the line BC, namely $x_0 = 0$, in $(0, v, -w)$. Similarly the other two points which are claimed to be collinear with $(0, v, -w)$ are found to be $(-u, 0, w)$ and $(u, -v, 0)$.

All three are on the line whose equation is

$$u^{-1}x_0 + v^{-1}x_1 + w^{-1}x_2 = 0 \, .$$

The converse of the theorem is proved similarly.

It will be noticed that the same proof applies in continuous projective geometry.

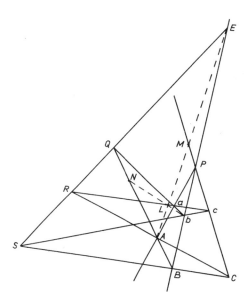

Fig. 3
The points E, M, N, L are not necessary for illustrating the
theorem of Desargues, but they will be used in the sequel.

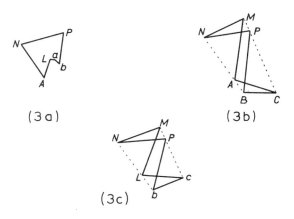

(3a)

(3b)

(3c)

The proof relies on commutativity of multiplication in the algebra from which the plane was derived, and does not apply if that algebra was a general $A[s]$ and not a $GF(s)$.

For example, in a finite projective plane derived from the $A[9]$ quoted in Chapter I, which is not the $GF(9)$, Desargues's theorem, does not hold.

Example

Consider the points $A(1, 0, 0)$, $B(0, 1, 0)$, $C(0, 0, 1)$; $P(2, 1, 1)$; $a(1, 1, 1)$, $b(2, 0, 1)$, $c(i, 2i, 1)$.

Then the lines AaP BbP CcP are, respectively $x_1 + 2x_2 = 0$, $x_0 + x_2 = 0$, $x_0 + x_1 = 0$.

AB and ab, i.e. $x_2 = 0$ and $x_0 + x_1 + x_2 = 0$ intersect in $(1, 2, 0)$, BC and bc, i.e. $x_0 = 0$ and $x_0 + x_1(2+i) + x_2 = 0$ intersect in $(1+i, 0, 1)$, AC and ac, i.e. $x_1 = 0$ and $x_0(1+i) + x_1(1+2i) + x_2 = 0$ intersect in $(1+i, 0, 1)$. The line $x_0 + x_1 + x_2(1+2i) = 0$ contains the first two intersections, but not the third.

A plane in which Desargues's theorem does not hold will be called *non-Desarguesian*.

In [62] Hannah Neumann has shown that there exist non-Desarguesian planes containing complete quadrangles with collinear diagonal points, and also quadrangles with non-collinear diagonal points. G. Pickert mentions in [66] his conjecture that it happens only in Desarguesian planes that the diagonal points of a complete quadrangle are either collinear in all quadrangles, or in none of them. Gleason has proved in [22] that a Fano-plane is always Desarguesian. The question whether a plane where no complete quadrangle has collinear diagonal points must be Desarguesian is yet unsolved, to the author's knowledge.

The theorem of Pappus

Pappus's theorem in ordinary projective geometry states that if A, B, C are three distinct points on a line, and a, b, c are three distinct points on another line, then the intersections of the pairs of lines Ab and aB, Ac and aC, Bc and bC are collinear (see Fig. 4).

It has been proved (see, e.g., [105]) that in a finite projective plane this theorem can be derived from that of Desargues. The converse is also true, as we shall now show, adapting [42] p.117 (see also [40]).

Consider, again, Fig. 3. The lines Aa, Bb and Cc pass through P. AB and ab meet in Q, and AC and ac meet in R. Let QR meet

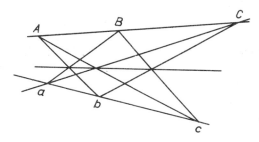

Fig. 4

BC in S, and let bc meet BC in S'. We want to prove that $S = S'$, assuming Pappus's theorem to hold.

Denote the intersection of Bb and RQ by E, and draw AE. Denote its intersection with ac by L, and that with Cc by M. Denote the intersection of Lb and AB by N. We prove our statement by a three-fold application of the theorem of Pappus.

First, we show that PN passes through R (Fig. 3a). In the hexagon $NALabP$ (PN is a line not shown in Fig. 3) NA and ab meet in Q, AL and Pb in E, hence PN and aL (i.e. ac) meet on EQ, i.e. in R, where AC and PN meet as well.

Next, we show that NM passes through S (Fig. 3b). In the hexagon $ACBPNM$ (NM is a new line) PN and AC meet in R, MA and BP in E, hence NM and BC meet on RE, i.e. in S.

Finally, consider Fig. 3c. In the hexagon $NPbcLM$, PN and Lc (i.e. aL) meet in R (see Fig. 3a), ML and bP meet in E, and hence NM and bc meet on RE. But NM and BC meet also on RE, i.e. in S (see Fig. 3b). Therefore, this is the point S' where BC and bc meet; thus $S = S'$.

The theorem of Pappus is a special case of Pascal's theorem about six points on a conic. Hence, instead of talking about a plane in which Pappus's theorem holds, some authors speak of "Pascalian planes", and it may be said that a finite projective plane which is non-Desarguesian is non-Pascalian, and vice versa.

Co-ordinates

We have seen that in a geometry derived from a Galois field Desargues's theorem holds, while this is not necessarily true when an algebra $A[s]$ was used for derivation.

34

We shall now show that a finite projective plane in which Desargues's theorem holds can always be derived from a Galois field. Hence Desarguesian planes with $s + 1$ points on a line exist only if s is the power of a prime and, in an abstract sense, there exists only one single Desarguesian plane for any given prime power s.

Let a finite projective Desarguesian plane be given with $s + 1$ points on each line. We introduce the elements of a $GF(s)$ as coordinates, adapting a method used by Hilbert in [41], pp. 79 ff.

To begin with, we attach elements of the $GF(s)$ to the points of two selected lines, say L_1 and L_2, which intersect in O (Fig. 5).

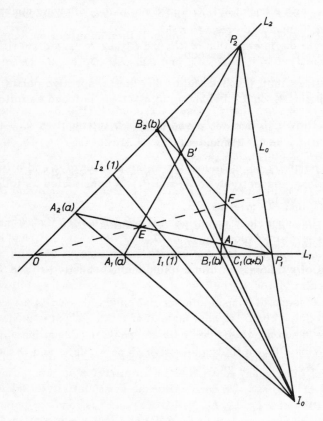

Fig. 5
This figure is also used for the proof that addition is commutative.

Attach to this point the element 0. Choose points on L_1 and on L_2, say l_1 and l_2 respectively, and attach to both the element 1. Then choose a third line, L_0 say, and denote its intersections with L_1 and L_2 respectively by P_1 and P_2, and its intersection with $l_1 l_2$ by l_0.

We attach to the $s - 2$ points on L_1 apart from 0, l_1 and P_1 the $s - 2$ elements of $GF(s)$ distinct from 0 and from 1. The elements to be attached to the points of L_2 are fixed by the rule that any point of L_2 has the same element attached to it as has its projection from l_0 on to L_1. (This is already true for l_1 and l_2.)

We now have to define the point C_1 on L_1 with the element $a + b$ attached to it, dependent on the point A_2 on L_2 with element **a**, and B_1 on L_1 with element **b**. This is done by the following construction.

Draw $A_2 P_1$ and $B_1 P_2$, and call their intersection A'.
Draw $A' l_0$; it intersects $O P_1$ in C_1.

It follows that, given **c** and **b**, and their respective points C_1 and B_1, the point A_2 on L_2 belonging to $a = c - b$ is found as follows:

Draw $C_1 l_0$ and $B_1 P_2$ and call their intersection A'.
Draw $P_1 A'$. It intersects L_2 in A_2.

The point D_2 on L_2 belonging to the product of **a** and **b** (where **a** and **b** belong, respectively, to A_2 and B_1) is constructed as follows (see Fig. 6 overleaf):

Draw $A_2 l_1$ and denote its intersection with L_0 by K.
Draw $B_1 K$; it intersects L_2 in D_2.

It follows that, given **d** and **b** and their respective points D_2 and B_1, the point A_2 on L_2 belonging to $a = db^{-1}$ is found as follows:

Draw $D_2 B_1$ and denote its intersection with L_0 by K.
Draw $l_1 K$; it intersects L_2 in A_2.

We must now show that as a consequence of the validity of the theorem of Desargues, addition and multiplication as just defined obey the rules of a $GF(s)$. We derive first the commutativity of addition.

Let the lines L_0, L_1, L_2, $A_1 A_2$, $B_1 B_2$ and $l_1 l_2$, and the points O, l_0, P_1 and P_2 be given as in Fig. 5. $A_2 P_1$ and $B_1 P_2$ intersect in A'. Draw $A_1 P_2$ and $B_2 P_1$ and let them intersect in B'. The line $l_0 A'$ meets L_1 in C_1, and the line $l_0 B'$ meets L_1 in C'. The element attached to

36

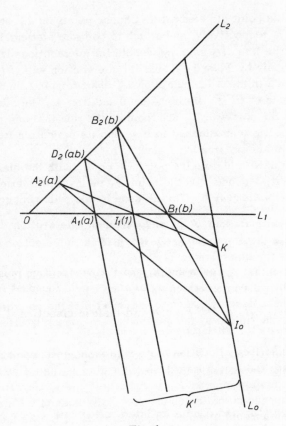

Fig. 6
This figure is also used for the proof that multiplication is commutative.

C_1 is $a+b$, and that attached to C' is $b+a$. If the commutative law holds, then C_1 and C' must be identical: I_0A' and I_0B' must meet L_1 in the same point. In other words, A', B' and I_0 must be collinear. This is the fact that we are going to prove, using Desargues's theorem.

Let A_1P_2 and A_2P_1 meet in E, and B_1P_2 and B_2P_1 meet in F. Applying the theorem to the triangles A_1A_2E and B_1B_2F, with their corresponding sides meeting on L_0, we see that O, E, and F are on the same line.

Hence, applying the theorem to the triangles OA_1A_2 and $FA'B'$ with their corresponding vertices on lines through E, we see that $A'B'$ and A_1A_2 intersect on a line through P_1 and P_2, i.e. in I_0.

The commutativity of multiplication can be proved as follows.

In Fig. 6, we turn to A_1 and B_2, instead of A_2 and B_1. Consider the two lines L_0 with the points K, I_0 and K', and L_1 with the points A_1, I_1 and B_1. Pappus's theorem shows then that B_2, D_2 and A_2 are collinear, i.e. the point with element ba is the same as that with element ab, namely D_2.

We are now ready to introduce co-ordinates for all points of the plane.

The point O has co-ordinates $(1, 0, 0)$. A point on L_1 has co-ordinates $(1, x_1, 0)$ and a point L_2 has co-ordinates $(1, 0, x_2)$, where x_1 and x_2 are the elements of $GF(s)$ belonging, respectively, to these points.

To find the co-ordinates of a point not on L_0, we draw PP_2 and note the co-ordinates $(1, y_1, 0)$ of its intersection with L_1, and draw PP_1 and note the co-ordinates $(1, 0, y_2)$ of its intersection with L_2. Then P has the co-ordinates $(1, y_1, y_2)$.

A point on L_0 has co-ordinates $(0, 1, y_2)$ where $(1, 1, y_2)$ are the co-ordinates of the intersection of $I_1 P_2$ and OP_0.

In all these cases we have to remember that the co-ordinates of any point can be multiplied by the same non-zero constant, without changing the point.

If the described allocation of co-ordinates seems complicated or even arbitrary, the reader may think of the line L_0 as a line "at infinity" and of lines intersecting on L_0 as "parallel" lines. He will then recognize that the co-ordinates are precisely those of a Cartesian plane, provided the elements attached to points on L_1 or L_2 are their Euclidean distances from O.

We have to show that these definitions are such that co-ordinates of all points on the same line satisfy a linear homogeneous equation. The co-ordinates of points on the line L_0 satisfy $x_0 = 0$, those on L_1 satisfy $x_2 = 0$, and those on L_2 satisfy $x_1 = 0$. The co-ordinates of points on a line through P_1 satisfy the equation $x_2 - k_1 x_0 = 0$, with an appropriate k_1, and the co-ordinates of points on a line through P_2 satisfy the equation $x_1 - k_2 x_0 = 0$ with an appropriate k_2.

We turn to a line L' through O, different from either L_1 or L_2 (see Fig. 7 overleaf).

Let the intersection of L' with $I_1 P_2$ be C, and that of $P_1 C$ with L_2 be D. Suppose the element attached to D is $\mathbf{a}$.

Consider now any point P' on L'. Its co-ordinates are $(1, m, n)$ and we denote the points belonging to the elements $\mathbf{m}$ on L_1 and $\mathbf{n}$ on L_2, i.e. the points $(1, m, 0)$ and $(1, 0, n)$ respectively by U and V.

38

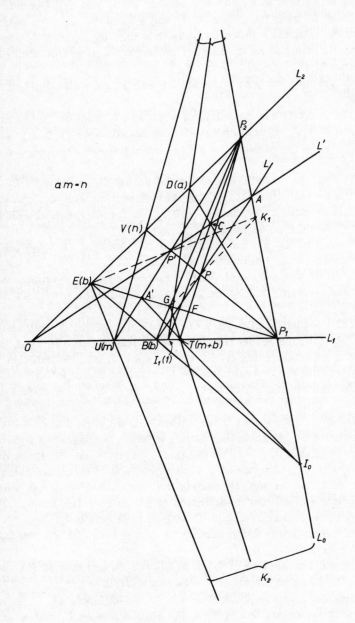

Fig. 7

Draw the lines I_1D and UV. The triangles I_1DC and UVP' have corresponding points connected by lines meeting in O. Hence corresponding sides meet on a line. It must be the line P_1P_2, i.e. L_0, and hence I_1D and UV meet on L_0. By the definition of multiplication this means that $am = n$, and therefore the co-ordinates of all points on L', a line through O, satisfy the equation $ax_1 - x_2 = 0$, where a is found by constructing D, as explained.

Now take a line L in general position, which intersects L_1 in B with element b, and L_0 in A. Through O, draw the line OA, which we denote by L' (Fig. 7).

Choose an arbitrary point on L, say P, and draw PP_1. Let it intersect L' in P' and L_2 in V. Draw PP_2 and let it intersect L_1 in T, and let P_2P' intersect L_1 in U.

The x_2 co-ordinate of P is the same as that of P', say n. Its x_1 co-ordinate is that of T, and we shall show that this is $m + b$, where m is the x_1 co-ordinate of P'. To this end, transfer b to L_2 via I_0. This gives the point E. To produce $m + b$, we draw EP_1 and UP_2. These lines intersect in A'.

Draw $A'I_0$. If this line intersects L_1 in T, then T will have its x_1 co-ordinate equal to $m + b$, according to our definition as will now be shown. For the proof we need more points and lines. Let F be the intersection of EP_1 and TP_2, and G that of EP_1 and BP_2.

Consider the triangles OEP' and BGP. Lines joining corresponding vertices meet in P, hence corresponding sides meet on a line through P_2 and A. Therefore, EP' and GP meet on L_0, say in K_1.

Now consider the triangles UEP' and TGP. Corresponding vertices lie on a line through P, hence corresponding sides meet on a line through K_1 and P_2, i.e. UE and TG meet on L_0, say in K_2.

Then, consider the triangles UEA' and GTB. Corresponding sides meet on L_0, so UG, ET, and $A'B$ meet in one point (not shown in (Fig. 7).

Finally, since corresponding vertices of the triangles GTA' and UEB are on lines through one point as has just been proved, $A'T$ and BE meet on the line K_2P_1, i.e. on L_0. They thus meet in I_0, as was to be proved.

We have now proved that the co-ordinates of an arbitrary point P on L satisfy the equation $x_1 = m + b$, where b is constant for all points on L, and $am = n$, n being the x_2 co-ordinate of P. Because P has co-ordinate $x_0 = 1$, we can write $ax_1 = x_2 + abx_0$, or $ax_1 - abx_0 - x_2 = 0$. This again is a linear homogeneous equation.

We have now seen that a Desarguesian plane can always be derived from a Galois field. Since such fields exist only when the order is a prime power, a projective plane with $s+1$ points on each line, where s is not a prime power, will have to be non-Desarguesian. It is not known whether such planes for s not a prime power exist.

We have not said, of course, nor would it be true, that a projective plane derived from a more general algebra cannot be Desarguesian. In fact, if it can be proved that apart from isomorphisms there exists only one finite projective plane with $s+1$ points on a line, and if s is the power of a prime, then this plane must be Desarguesian.

It has been shown that this is the case for $s = 2, 3, 4, 5, 7$ (see [27]) and for 8 (see [34]). For any s which equals a prime, only Desarguesian planes are known. On the other hand, it has been shown by A. A. Albert in [2] that for $s = p^r$ where p is odd and r at least 2, several non-Desarguesian planes exist. They exist also for $s = 2^{2^r}$ where r is at least 2. No plane exists for $s = 6$ (see [101] or [12]). For $s > 8$ the precise number of planes is unknown.

To give an idea of the way in which the uniqueness of a projective plane can be proved, we show here that if there are five points on each line, then the plane must be a $PG(2, 4)$. We prove, in fact, that every such plane is Pascalian, and we have already mentioned that it must then be Desarguesian as well.

In Fig. 4, consider the lines DE, DF and EF. We prove that they are, in fact, identical.

None of them intersects OA in either A or B or C. Apart from these points, there is O on OA, and only one more point. But if DE, EF, and DF were different lines, then they would produce three more intersections with OA, which is impossible. (This proof is taken from [105]).

It is also of interest to determine necessary conditions for a projective plane P with $n+1$ points on a line to contain a projective plane P' with $m+1$ points on a line, where $m < n$.

Take a line in P' and a point on it, say Q, which is in P, but not in P'. Join Q to the m^2 points of P' not on that line. The resulting m^2 lines are all different, since otherwise they would join two points of P', and hence Q, being the intersection of two lines in P', would also lie in it. Thus we have m^2+1 lines through Q, and therefore $n+1 \geqslant m^2+1$.

More precisely, if n exceeds m^2, then $n \geqslant m^2+m$. This is so, because then there is a line through Q with no point in P', and its intersections with the m^2+m+1 lines of P' are all distinct, since otherwise the common intersection would be a point of P'. Therefore $n+1 \geqslant m^2+m+1$. Thus a necessary condition for P' to be contained in P is that either $n = m^2$, or $n \geqslant m^2+m$. However, this condition is not sufficient.

Conics

We can go further in introducing concepts analogous to those with which we are familiar in continuous geometry, and we shall do this for Desarguesian planes, derived from a $GF(s)$.

Let the equations of two lines be —
$$a_{01}x_0 + a_{11}x_1 + a_{21}x_2 = 0 \quad \text{and} \quad a_{02}x_0 + a_{12}x_1 + a_{22}x_2 = 0.$$

Let the intersection of these two lines be P. All lines through P form a pencil, and each line in this pencil has an equation of the form

$$(ma_{01} + na_{02})x_0 + (ma_{11} + na_{12})x_1 + (ma_{21} + na_{22})x_2 = 0,$$

where m and n are elements of $GF(s)$ and not simultaneously 0. There are altogether $s+1$ lines in the pencil, namely the two original lines corresponding to $n = 0$ and $m = 0$ respectively, and those corresponding to $s - 1$ different ratios of m/n with $m \neq 0$ and $n \neq 0$.

Let a second pencil be defined by
$$(mb_{01} + nb_{02})x_0 + (mb_{11} + nb_{12})x_1 + (mb_{21} + nb_{22})x_2 = 0.$$

We define a projective correspondence between the lines of the two pencils by letting a line of the first, given by a pair, m, n, correspond to a line of the second pencil which corresponds to the same pair. Two corresponding lines meet in a point, and the co-ordinates of all the $s + 1$ points so obtained satisfy the equation which we find by eliminating m and n from the equations of the two pencils, i.e. the equation

$$(a_{01}x_0 + a_{11}x_1 + a_{21}x_2)(b_{02}x_0 + b_{12}x_1 + b_{22}x_2) -$$
$$(a_{02}x_0 + a_{12}x_1 + a_{22}x_2)(b_{01}x_0 + b_{11}x_1 + b_{21}x_2) = 0.$$

The left-hand side is a quadratic form and the set of points whose co-ordinates satisfy the equation is called a *conic*. It is called *degenerate* if the left-hand side is the product of two linear forms. It consists then of the $2s + 1$ points of two intersecting lines. A *non-degenerate* conic contains the $s + 1$ points which appear in its construction. For the proof that the equation is satisfied by precisely $s+1$ points we refer the reader to [68].

In [83] Segre has proved the correctness of a conjecture in [44], that any set of $s + 1$ points in a $PG(2, s)$ such that no three of them are on a line is a non-degenerate conic.

A line which has precisely one point in common with a conic is called a *tangent* of it; a line that has two points in common is a *secant*. Because the equation of a conic is quadratic, it cannot have more than two points in common with any line.

Take one point of a non-degenerate conic, and connect it by lines to the other s points. The resulting lines are secants and the remaining one of the $s + 1$ lines through that point must be a tangent.

We shall now exhibit a difference between conics in Desarguesian planes of odd and of even order.

First, let the order s of the plane be odd. Take a point not on the conic, P say, and connect it by lines with all the $s + 1$ points on the conic. These lines are either secants or tangents. Since each secant accounts for two points of the conic, and since $s + 1$ is even, the number of tangents among these lines must also be even. But we can prove that it cannot be more than 2. For suppose P were a point with more than two tangents of the conic passing through it. Call three of them t_0, t_1 and t_2 and let the point of the conic on t_0 be P_0. Of the points of intersection of t_0 with the other s tangents two will coincide with P. If all the other $s - 2$ intersections of t_0 with other tangents are distinct, then there will be $(s+1) - (s-2) - 2 = 1$ point through which there is no tangent except t_0. (The last term, -2, counts P and P_0.)

If, on the other hand, the other $s - 2$ intersections are not all distinct, then there will be even more points on t_0 through which no other tangent passes. But this is impossible, because no point outside the conic can have an odd number of tangents through it. We have proved

Theorem 2.1. Through a point outside a conic in a Desarguesian plane of odd order there pass none or two tangents of the conic.

Now let the order of the plane be even. We may assume, without loss of generality, that three points on a non-degenerate conic are $A(1, 0, 0)$, $B(0, 1, 0)$, $C(0, 0, 1)$. The tangents through these three points are, respectively, $x_1 - k_0 x_2 = 0$, $x_2 - k_1 x_0 = 0$, and $x_0 - k_2 x_1 = 0$, for if, for instance, there were another point of the conic on the first-mentioned line, then this point would be on a line through B and C.

Let $P(t_0, t_1, t_2)$ be another point of the conic. Again, none of the t_i can be 0, because if it were, then P would be on a line through two of the points A, B, and C. We can therefore write the equations:

$$x_1 - (t_1/t_2)x_2 = 0 \text{ for } PA, \qquad x_2 - (t_2/t_0)x_0 = 0 \text{ for } PB \quad \text{and}$$

$$x_0 - (t_0/t_1)x_1 = 0 \text{ for } PC.$$

Consider the equation for PA. As we choose for P the various points of the conic, leaving out A, B and C, the ratio t_1/t_2 takes all the values of the elements of $GF(s)$ apart from 0 and k_0. Since (see Chapter I) $(x - x_1)\ldots(x - x_{s-1}) = x^{s-1} - 1$, where the x_i are the non-zero elements of $GF(s)$, the product of all non-zero elements is $(-1)^s$. Therefore, if we multiply the product of the $s-2$ values which t_1/t_2 assumes by k_0, we obtain $(-1)^s = 1$ (because s is even).

We write $k_0 \Pi(t_1/t_2) = 1$ where the product extends over all points of the conic except A, B and C. Similarly, we have $k_1 \Pi(t_2/t_1) = 1$ and $k_2 \Pi(t_0/t_1) = 1$.

For all points apart from A, B and C we have $t_1 t_2 t_0 / t_2 t_0 t_1 = 1$, and hence multiplying the three expressions of the products above we have $k_0 . k_1 . k_2 = 1$. Therefore the points $(1, k_0 k_1, k_1)$, $(k_2, 1, k_1 k_2)$ and $(k_0 k_2, k_0, 1)$ are identical. The three tangents in A, B and C pass through this point, and because these points were arbitrary, any three tangents meet in the same point. This proves

Theorem 2.2   In a finite projective plane of even order the tangents of a non-degenerate conic meet in the same point. We call this point the *nucleus* of the conic. The proof of the theorem is due to B. Segre (see [84]).

No secant passes through the nucleus, because there are only $s+1$ lines through it altogether. Thus the points of the conic and its nucleus are $s+2$ points, no three of them on the same line.

If the order s of a finite projective plane is odd, then it is not possible to find a set of $s+2$ points such that no three of them are on the same line. This is so, because if there were such a set, then every one of the $s+1$ lines through one of its points would pass through one of the other points, and hence there would not be any line in the plane which passes through only one of the points (a tangent). But this is impossible. For consider the $s+1$ lines through one of the points not in the set. These lines may have none, one, or two points in common with the set. Those of the first and third types account for an even

44

number of points. If s and therefore $s+2$ are odd, then there must be points of the set left through which a tangent passes. We would thus reach a contradiction. This proves

Theorem 2.3 In a finite projective plane of odd order s there will be in any set of $s+2$ points at least one subset of three points which lie on the same line.

In a plane of even order 2^n it can be shown that in those cases where a value k exists such that $2^n-1 = \binom{k-1}{2}$ (for instance $k = 3$, $n = 1$, or $k = 4$, $n = 2$, or $k = 7$, $n = 4$),* any k points $P_1 \ldots P_k$ of a set of 2^n+2 points with no three points on a line determine uniquely all its points. Take one of the k points, say P_1. Of the 2^n+1 lines through P_1, $k-1$ are the lines $P_1P_2, \ldots, P_1P_k$. These lines do not contain any other point of the set. Each of the remaining 2^n+2-k lines through P_1 meets the $\binom{k-1}{2}$ lines $P_2P_3, \ldots, P_2P_k, \ldots, P_{k-1}P_k$ in $\binom{k-1}{2}$ distinct points. None of these can belong to the set; however, on each of these lines there remain $2^n - \binom{k-1}{2} = 1$ further point, apart from P_1 and those intersections. These further points make up 2^n+2 points, and these must be the points of the set. (The case for $k = 4$, $n = 2$ is contained in [85].)

A non-degenerate conic in a plane of order s consists of $s+1$ points. If s exceeds 4, then 5 points are sufficient to determine it. The proof of this is analogous to the proof in continuous geometry and depends on the fact that a quadratic form in three variables has six coefficients, determining five ratios.

The first point can be selected in s^2+s+1 different ways, the second in s^2+s ways. The third must not be on the line defined by the two points already chosen, so that s^2 possible choices remain. The fourth and fifth can, respectively, be chosen in $(s-1)^2$ and in $(s-2)(s-3)$ ways. The five points might have been selected in any order, so that there are

$$(s^2+s+1)(s^2+s)s^2(s-1)^2(s-2)(s-3)/(s+1)s(s-1)(s-2)(s-3) = s^5-s^2$$

different non-degenerate conics in the plane. Of these there will be

$$(s^2+s)s^2(s-1)^2(s-2)(s-3)/s(s-1)(s-2)(s-3) = s^2(s^2-1)$$

---

* U.V. Satyarayana has shown in [82] that there are infinitely many natural numbers t such that $2^{2t}-1$ is not a triangular number.

through a given point and

$$s^2(s-1)^2(s-2)(s-3)/(s-1)(s-2)(s-3) = s^2(s-1)$$

through two given points.

It is immediately seen that this is also true if $s = 2$ or 3.

Euclidean planes

We have called the finite planes projective, because of their analogy with the better known continuous projective planes. An analogue of a continuous Euclidean plane can be constructed by omitting, from a finite projective plane, one line and all the points on it. There will, then, be s^2 points and s^2+s lines left, with a line through any pair of points, and s points on every line. It is not true any more that any two lines meet in a point. Two lines which have no point in common might be called *parallel*. The Euclidean plane derived in this way from a $PG(2,s)$ is denoted by $EG(2,s)$.

An $EG(2,s)$ is a partial plane according to the definition in [28]. There Marshall Hall, Jr. defines a finite partial plane as a set of points and lines (subsets of points) such that two distinct lines intersect in at most one point, and two points lie on at most one line.

Example

Consider Fig. 1 and omit the line AFB, with its three points. Then the lines DC and EG are parallel, and so is the pair of lines CG and DE, and the pair DG and CE. This $EG(2,2)$ is a quadrangle with its six sides.

Consider in a $PG(2,s)$ those conics which have a common tangent. There are as many conics through one point of it, and having that tangent, as there are conics through two points, and thus we have altogether $s^2(s-1)(s+1) = s^2(s^2-1)$ conics with a common tangent. If we omit the latter and all its points from the $PG(2,s)$, then in the remaining $EG(2,s)$ these conics will have s points each. We call them *parabolae*, and there will be $s(s^2-1)$ of them through each of the s^2 points of the $EG(2,s)$.

Conics which have two points in common with the omitted line are *hyperbolae*. They contain $s-1$ points each, and there are $\binom{s+1}{2} s^2(s-1) = \frac{1}{2}(s^2-1)s^3$ of them. It follows that there are

$$s^5 - s^2 - s^2(s^2-1) - \frac{1}{2}(s^5-s^3) = \frac{1}{2}(s^5-s^3) = \frac{1}{2}s^3(s-1)^2$$

conics with $s+1$ points each (*ellipses*).

46

We do not discuss here Euclidean planes in greater detail, because we shall meet a generalization of such planes in the next chapter.

EXERCISES

(1) In a $PG(2,4)$, consider the quadrangle

$$A(1,1,1+y), \quad B(0,1,y), \quad C(1,1,y), \quad D(1,1+y,y)$$

(For notation, see Exercise I.3).

Find its diagonal points, and prove that they are collinear.

(2) In a $PG(2,5)$, consider the triangles

$$A(3,0,2), \quad B(1,4,0), \quad C(1,3,2)$$

and $\qquad a(1,0,4), \quad b(2,3,0), \quad c(2,4,1)$.

Determine the remaining elements (points and lines) of the Desarguesian configuration of ten points and ten lines which can be derived from them.

(3) There are six points in a $PG(2,4)$, no three of which are on the same line. Four of them are A, B, C, and D of Exercise (1) above. Find the other two. (Cf. page 44.)

(4) Choose a non-degenerate conic in a $PG(2,4)$ and determine its tangents and its nucleus.

(5) Find a non-degenerate conic in the $PG(2,5)$ and determine all its tangents. Show that there is no nucleus (i.e. the tangents are not all concurrent).

(6) Find the equation of the conic constructed in the solution of Exercise (4).

CHAPTER III

FINITE SPACES OF HIGHER DIMENSIONS

Definitions

A finite projective space is a finite set of points, with subsets called *lines*, subject to the following rules (cf. [104]):

(1) A line contains at least three points.

(2) There is precisely one line through any pair of distinct points.

These two rules appear also in the rules, or the consequences of the rules, for finite projective planes. The third rule for planes does not apply in more general spaces (there exist skew lines, just as in continuous three-dimensional geometry). However, we want finite projective spaces to contain finite projective planes, and this is achieved by the following rule (which, in a finite plane, is equivalent to a rule given in Chapter II):

(3) A line which intersects two sides of a triangle intersects the third line as well.

A k-space is defined as follows:

A 0-space is a point. If $A_0 \ldots A_k$ are points not all in the same $(k-1)$-space, then all points collinear with A_0 and any point in the $(k-1)$-space defined by $A_1 \ldots A_k$ form a k-space. Thus a line is a 1-space, and all other spaces are defined recurrently. The existence of an m-space $(m > 1)$ is ensured by the following further rules:

(4) If $k < m$, then not all points considered are in the same k-space.

(5) There exists no $(m + 1)$-space in the set of points considered.

It follows that a 2-space is a finite projective plane. We shall say that an m-space has m dimensions, and if we refer to a k-space as a subspace of a space of higher dimensions we call it a *k-flat*. An $(m - 1)$-flat in a space of m dimensions will sometimes be called a *hyperplane*.

The theorem of Desargues

Using these rules and definitions, it can be proved that in any 2-flat

48

(plane) in a space of at least three dimensions the theorem of Desargues (see Chapter II) is always valid. This theorem can only fail to be true in planes that cannot be embedded in a space of more than three dimensions.

The proof depends on the fact that within a space of m dimensions an a-flat and a b-flat have a c-flat in common, where $c = a + b - m$, provided this number is not negative. (The demonstration of this follows from the rules, but the proof is too lengthy to be given here. It was first given in a manuscript written by members of a class of O. Veblen and deposited in the mathematical library of the University of Chicago.) In what follows we shall need only the fact that in a 3-space a line and a plane have at least one point, and two planes have at least one line, in common. Moreover, two intersecting lines define a plane in which they lie.

Consider two triangles, ABC and abc, in the same plane (Fig. 8), and such that Aa, Bb and Cc pass through the same point, P, say.

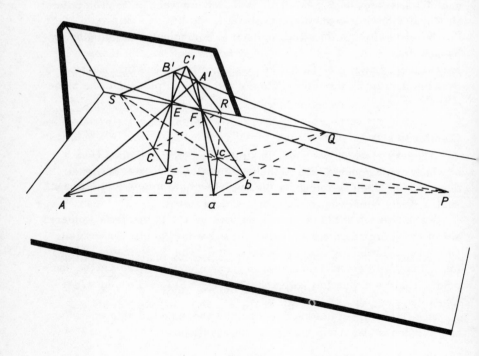

Fig. 8

Let the points mentioned be all different. (There exist generalizations of this situation in which the theorem is still true, but we do not discuss them here.)

Through P draw a line not on the plane common to the two triangles, and take on this line two points E and F, say, different from P and from one another. Both the line EA and Fa lie in the plane defined by EPa (which is also the plane EPA, FPA, or FPa). Hence they have a point in common, say A'. Similarly EB and Fb intersect in B' and EC and Fc intersect in C'. The points A', B' and C' are not collinear, because if they were, then they would all three be, with E, in a plane different from that of the two original triangles, but containing A, B, and C, so that these last three points would be collinear and would not form a triangle.

Now consider the points A' and B'. They are, respectively, on EA and on EB, so that A, B, A', B' and E are on the same plane. Hence the lines $A'B'$ and AB intersect in a point, say Q. Similarly $A'C'$ and AC intersect in R, and $B'C'$ and BC intersect in S. The points Q, R and S are all in the original plane.

In precisely the same way, replacing E by F, we find that ab passes through Q, ac through R, and bc through S. Our theorem is proved if it is seen that Q, R and S are collinear, and this must be so, because they lie in the intersection of the original plane and the plane $A'B'C'$.

Co-ordinates

The way in which we have constructed projective planes from algebras can be generalized to the construction of spaces of higher dimensions. Since Desargues's theorem must hold in every plane, we shall restrict ourselves to the use of Galois fields.

We define a point in m-space, denoted by $PG\,(m, s)$, as an ordered set of $m + 1$ co-ordinates $(x_0 ,..., x_m)$, where the x_i are elements of $GF(s)$ $(s \geqslant 2)$, and not all simultaneously 0. The co-ordinates $(ax_0,..., ax_m)$, where a is a non-zero element of $GF(s)$, define the same point. There are therefore altogether $(s^{m+1} - 1)/(s - 1)$ points in a $PG(m, s)$.

A k-flat is the set of all those points whose co-ordinates satisfy $m - k$ independent linear homogeneous equations

$$a_{10}x_0 + \cdots + a_{1m}x_m = 0$$

$$- \ - \ - \ - \ - \ - \ -$$

$$a_{m-k0}x_0 + \cdots + a_{m-km}x_m = 0$$

with coefficients from $GF(s)$, not all simultaneously 0 within the same equation. Alternatively, a k-flat consists of all those points with coordinates

$$(a_0 x_{00} + \cdots + a_k x_{k0} , \ldots, a_0 x_{0m} + \cdots + a_k x_{km})$$

with all elements from $GF(s)$, where the a_i are not simultaneously 0, and the $k+1$ points $(x_{00} \cdots x_{0m}), \ldots, (x_{k0}, \ldots, x_{km})$ are linearly independent, i.e. the matrix

$$\begin{pmatrix} x_{00} \cdots x_{0m} \\ - \ - \ - \ - \\ x_{k0} \cdots x_{km} \end{pmatrix}$$

has rank $k+1$. (Such points exist, e.g. $(1, 0, \ldots, 0)$, $(0, 1, 0, \ldots, 0)$ ... $(0, \ldots, 0, 1, 0, \ldots, 0)$.) It follows that $(a_0 x_{0i} + \cdots + a_k x_{ki})$ cannot simultaneously be 0 for all i, because this would mean that $a_0 = \ldots = a_k = 0$. The equivalence of the two definitions follows from algebraic rules in a way similar to the derivation of the analogous fact in a $PG(2, s)$.

The number of points in a k-flat can be found as follows. There are $(s^{k+1} - 1)$ sets of elements $a_0, \ldots, a_k$, excluding the set $(0, \ldots, 0)$. But not all of these represent different points. Two points $(a_0 x_{00} + \cdots + a_k x_{k0}, \ldots, a_0 x_{0m} + \cdots + a_k x_{km})$ and $(b_0 x_{00} + \cdots + b_k x_{k0}, \ldots, b_0 x_{0m} + \cdots + b_k x_{km})$ are equal if there exists a c different from 0, such that $b_0 x_{0i} + \cdots + b_k x_{ki} = c(a_0 x_{0i} + \cdots + a_k x_{ki})$ for $i = 0, 1, \ldots, m$.

By virtue of our assumption about the independence of the $k+1$ points this is so if and only if $b_0 - ca_0 = \ldots = b_k - ca_k = 0$. Hence, given $a_0, \ldots, a_k$, there are $s-1$ sets of $b_0, \ldots, b_k$, corresponding to the non-zero values of c, which produce the same point. Therefore the k-flat contains altogether $(s^{k+1} - 1)/(s-1)$ points. There are $s+1$ points on a line and $s^2 + s + 1$ on a plane.

We show now that, given a k-flat, any $k+1$ independent points can serve as a basis. Indeed, if the points

$$P_0 = (a_{00}x_{00} + \cdots + a_{0k}x_{k0}, \ldots, a_{00}x_{0m} + \cdots + a_{0k}x_{km})$$

$$- \ - \ - \ - \ - \ - \ -$$

$$P_k = (a_{k0}x_{00} + \cdots + a_{kk}x_{k0}, \ldots, a_{k0}x_{0m} + \cdots + a_{kk}x_{km})$$

are independent, then the product of the matrices

$$\begin{pmatrix} a_{00} \cdots a_{0k} \\ - - - \\ a_{k0} \cdots a_{kk} \end{pmatrix} \text{ and } \begin{pmatrix} x_{00} \cdots x_{0m} \\ - - - \\ x_{k0} \cdots x_{km} \end{pmatrix}$$

has rank $k + 1$, so that the first-mentioned matrix cannot be singular. Hence the co-ordinates of any other point, say $(b_0 x_{00} + \ldots + b_k x_{k0}, \ldots, b_0 x_{0m} + \ldots + b_k x_{km})$, where the b_j are not all 0, can be computed as a linear combination of those of the given $k + 1$ points.

We must prove that a $PG(m, s)$ satisfies the five rules for a finite projective space.

(1) is assured because of $s \geq 2$.

(2) If two points, $(y_0, \ldots, y_m)$ and $(z_0, \ldots, z_m)$ are given, then the unique line through them is given by $(a y_0 + b z_0, \ldots, a y_m + b z_m)$, where a and b are elements from $GF(s)$.

(3) We may assume, without loss of generality, that the triangle is that of the points $A(1, 0, \ldots, 0)$, $B(0, 1, 0, \ldots, 0)$ and $C(0, 0, 1, 0, 0)$. A point on AB has co-ordinates $(a, b, 0, \ldots, 0)$ and one on BC has co-ordinates $(0, c, d, 0, \ldots, 0)$. A line through these two points will meet the line AC in the point $(ac, 0, -bd, 0, \ldots, 0)$.

(4) The points $(1, 0, \ldots, 0)$, $(0, 1, \ldots, 0), \ldots, (0, \ldots, 0, 1)$ are not all in the same k-space.

(5) With $m + 1$ co-ordinates, there exists obviously no space of higher dimension than m.

It can also be shown that co-ordinates can be introduced in any space satisfying the five rules. The proof is analogous to that for planes in Chapter II, and we shall not be concerned with it here (cf. [104] p. 246).

Duality

The double description of a k-flat, as an intersection of hyperplanes and as a combination of points, leads to the concept of duality. If a k-flat is defined as an intersection of $m - k$ linearly independent hyperplanes, then the coefficients in any of the latter can be considered as its homogeneous co-ordinates, and they can be treated in the same way as point co-ordinates. Then a point is dual to a hyperplane, a k-flat to a $(m - k - 1)$-flat, the k-flats in a t-flat are dual to the configuration of the $(m - k - 1)$-flats through a $(m - t - 1)$-flat, and these relationships can be used for enumeration. All enumerations remain valid if we make these replacements in the dimensions, and replace the phrase "in a flat" by "through a flat" and vice versa.

We introduce now a relationship between points and hyperplanes, determined by a bilinear form $\sum_i \sum_j a_{ij} x_i y_j$, where the a_{ij}, x_i and y_j are elements of $GF(s)$. Let a point $(y_0, \ldots, y_m)$ correspond to the hyperplane $\sum_i (\sum_j a_{ij} y_j) x_i = 0$, which contains all those points whose co-ordinates $(x_0, \ldots, x_m)$ satisfy the equation.

In particular, assume that the matrix of coefficients (a_{ij}) is skew-symmetric, i.e. $a_{ij} = -a_{ji}$, and hence $a_{ii} = 0$ for all i and j. Then $\sum_i \sum_j a_{ij} x_i x_j = 0$, so that every hyperplane contains the point which corresponds to it. Let the hyperplane to which $(y_0, \ldots, y_m)$ corresponds contain also the point $(z_0, \ldots, z_m)$, so that $\sum_i \sum_j a_{ij} y_j z_i = 0$. This can be written, after exchanging subscripts,

$$\sum_i \sum_j a_{ji} y_i z_j = -\sum_i \sum_j a_{ij} z_j y_i = 0$$

so that the hyperplane to which $(z_0, \ldots, z_m)$ corresponds also contains the point $(y_0, \ldots, y_m)$.

If m is even, then the skew symmetric matrix (a_{ij}) is singular. There will then be one point that does not correspond to any hyperplane, and vice versa (see Exercises).

Enumeration

Let us determine the number of k-flats in a $PG(m,s)$, when $k \leqslant m$. First, we find the number of sets of $k + 1$ independent points in a $PG(m, s)$. The first point can be selected in $(s^{m+1}-1)/(s-1)$ different ways. The second is then any out of $s^m + s^{m-1} + \ldots + s = s(s^m-1)/(s-1)$ points. The third point must be one not on the line through the first two, so that $s + 1$ points are thereby excluded from being chosen, which leaves $s^m + \ldots + s^2 = s^2(s^{m-1}-1)/(s-1)$. Continuing, we shall have for the $(k + 1)$th choice $s^k(s^{m-k+1}-1)/(s-1)$ possibilities.

Altogether, we thus have

$$\frac{s.s^2 \ldots s^k (s^{m+1}-1)(s^m-1)\ldots(s^{m-k+1}-1)}{(s-1)^{k+1}}$$

possibilities of choosing $k + 1$ independent points.

We have seen that any $k + 1$ independent points out of a k-flat define the latter. Therefore, in order to find the number of different

k-flats, we must divide the number just found by that of different sets of $k + 1$ points in one of the flats. Repeating the previous enumeration, but replacing m by k, we obtain

$$\frac{s.s^2 \ldots s^k (s^{k+1} - 1)(s^k - 1) \ldots (s - 1)}{(s - 1)^{k+1}}$$

The number of k-flats within a $PG(m, s)$ or also within an m-flat of a space of higher dimension is therefore

$$\frac{(s^{m+1} - 1)(s^m - 1) \ldots (s^{m-k+1} - 1)}{(s^{k+1} - 1)(s^k - 1) \ldots (s - 1)} = V(m, k; s), \text{ say.}$$

For $m = k$, we have $V(m, m; s) = 1$, as it must be. For $m \neq k$, we have $V(m, k; s) = V(m, m - k - 1; s)$. The equality of these two values can be shown by cross-multiplication. In particular, we have $V(m, m - 1; s) = V(m, 0; s) = (s^{m+1} - 1)/(s - 1)$; this is the number of hyperplanes and also that of points in a $PG(m, s)$.

The following tables of $V(m, k; s)$ for $s = 2, 3, 4, 5$ were worked out from the recurrence formula $V(m, k; s) = V(m, k - 1; s) \dfrac{s^{m-k+1} - 1}{s^{k+1} - 1}$.

The arrangement is as follows:—

$V(1, 0)$	$V(1, 1)$				
$V(2, 0)$	$V(2, 1)$	$V(2, 2)$			
$V(3, 0)$	$V(3, 1)$	$V(3, 2)$	$V(3, 3)$		
$V(4, 0)$	$V(4, 1)$	$V(4, 2)$	$V(4, 3)$	$V(4, 4)$	
$V(5, 0)$	$V(5, 1)$	$V(5, 2)$	$V(5, 3)$	$V(5, 4)$	$V(5, 5)$

$$s = 2$$

3	1				
7	7	1			
15	35	15	1		
31	155	155	31	1	
63	651	1395	651	63	1

$$s = 3$$

4	1				
13	13	1			
40	130	40	1		
121	1210	1210	121	1	
364	11011	33880	11011	364	1

$$s = 4$$

5	1				
21	21	1			
85	357	85	1		
341	5797	5797	341	1	
1365	93093	376805	93093	1365	1

$$s = 5$$

6	1				
31	31	1			
156	806	156	1		
781	20306	20306	781	1	
3906	508431	2558556	508431	3906	1

We also wish to know the number of t-flats within a $PG(m, s)$ which contain a given k-flat $(m \geqslant t \geqslant k)$. By duality this is the number of $(m-t-1)$-flats contained in a $(m-k-1)$-flat, namely $V(m-k-1, m-t-1; s)$ which is, if $t \neq k$, also $V(m-k-1, t-k-1; s)$. For instance, there are $V(m-1, 0, s)$ lines of the $PG(m, s)$ through a given point, $V(m-2, 0; s)$ different hyperplanes through a given line, and $V(m-1, 0; s)$ hyperplanes through a point, equal to the number of points on a hyperplane.

Cycles of flats; difference sets

We have seen that in a $GF(s^{m+1})$, where s is the power of a prime, all powers of an element x can be represented as polynomials in x of a degree not larger than m, with coefficients from $GF(s)$. If $x^i = a_0 x^m + \ldots + a_m$, then we may consider x^i as representing a point in a $PG(m, s)$ with co-ordinates $(a_0 \ldots a_m)$.

A k-flat through $(k + 1)$ independent points represented by x^{b_0}, ..., x^{b_k}, will contain all points represented by $\sum_{i=0}^{k} a_i x^{b_i}$, where the a_i are elements of $GF(s)$, not all 0.

Consider now the points $\sum_{i=0}^{k} a_i x^{b_i+c}$, $c = 0, 1, \ldots, v-1$, where $v = (s^{m+1}-1)/(s-1)$. They too are k-flats, because if x^{b_i+c} were a linear combination of the other powers x^{b_j+c}, then this would be true of x^{b_i} and the x^{b_j} as well. We denote the k-flat with given c by S_c, and we have $S_v = S_0$, because x^v (x being an element of $GF(s^{m+1})$) is again an element of $GF(s)$.

Let j be the smallest value for which $S_j = S_0$. Then $S_0 = S_j = S_{2j} = \ldots = S_v$ and hence j is a factor of v, say $v = rj$. We call j the *cycle* of S_0.

If x^{d_0} is a point of the k-flat S_0, then so are the points with exponents

$$d_0 \qquad d_0 + j \quad . \quad . \quad . \quad d_0 + (r-1)j$$

because $S_{hj} = S_0$ for $h = 0, 1, \ldots, r-1$. If there are further points on S_0, they can be written (again indicating only the exponents)

$$d_1 \qquad d_1 + j \quad . \quad . \quad . \quad d_1 + (r-1)j$$

$$- \quad - \quad - \quad -$$

$$d_{t-1} \qquad d_{t-1}+j \quad . \quad . \quad . \quad d_{t-1}+(r-1)j,$$

where $d_{i_1} - d_{i_2}$ is not divisible by j.

The number of all these distinct points is $rt = (s^{k+1}-1)/(s-1)$. Therefore, if $j < v$, and therefore $r > 1$, then $v = rj$ and rt have a common factor, namely r.

It follows that if $rj = (s^{m+1}-1)/(s-1)$ and $rt = (s^{k+1}-1)/(s-1)$ are relatively prime, then $r = 1$, $j = v$, and all k-flats have cycle v. This is the case for $k = m-1$ (because $s^m + s^{m-1} + \ldots + s + 1$ and $s^{m-1} + \ldots + s + 1$ cannot have a common factor), and it is the case for $k = 1$ when $(s^{m+1}-1)/(s-1)$ and $s+1$ are relatively prime, i.e. when m is even. For instance, every Desarguesian plane is cyclic (i.e. has cycle v). On the other hand, no cyclic plane which is not Desarguesian is known (see [29]).

It is proved in [71] that even if not all k-flats for given k have cycle v, there are some that have.

Example

$PG(3,2)$. 15 points, 35 lines, 15 planes. $s^{m+1} = 16$.

We can express the powers of a mark x as polynomials and establish thereby a correspondence between the powers and the co-ordinates of the points. We use here the irreducible equation $x^4 + x^3 + 1 = 0$ and obtain:

A (0001)…. x^0	F (0110)…. x^{13}	K (1011)…. x^5
B (0010)…. x^1	G (0111)…. x^7	L (1100)…. x^{14}
C (0011)…. x^{12}	H (1000)…. x^3	M (1101)…. x^{11}
D (0100)…. x^2	I (1001)…. x^4	N (1110)…. x^8
E (0101)…. x^9	J (1010)…. x^{10}	O (1111)…. x^6

The plane $a_0x^0 + a_1x^1 + a_2x^2$, where the a_i are elements of $GF(2)$, is the same as the plane $x_0 = 0$. It contains the points $A = x^0$, $B = x^1$, $C = x^{12}$, $D = x^2$, $E = x^9$, $F = x^{13}$, and $G = x^7$. It has a cycle 15, as has any other (hyper)plane. The plane $a_0x^1 + a_1x^2 + a_2x^3$, which is the plane $x_3 = 0$, contains $BDFHJLN$, etc.

The line $a_0x^2 + a_1x^{12}$, i.e. the line DCG, has cycle 5. We obtain the lines DCG, HFN, ILE, KAJ, OBM, and then GDC, which is again the line DCG, from which we started. On the other hand, the lines ABC and ADE both have cycle 15. Thus we account for all the $5 + 15 + 15 = 35$ lines.

Denote the number of points on a hyperplane, $(s^m - 1)/(s-1)$, by t and the points themselves by $x^{d1}, ..., x^{dt}$. Since a hyperplane has cycle $v = (s^{m+1} - 1)/(s-1)$, the several hyperplanes of the $PG(m, s)$ contain the points indicated by the exponents of x, as follows:

First hyperplane….	d_1	d_2	d_t
Second hyperplane….	$d_1 + 1$	$d_2 + 1$	$d_t + 1$
vth hyperplane….	$d_1 + v - 1$	$d_2 + v - 1$	$d_t + v - 1$

All values $0, 1, ..., v-1$ appear in this table t times, because there are t hyperplanes through each point.

We look now for those rows which contain a particular value, say 0. For this value to appear (as an exponent), we must have $d_i + h = 0$ ($i = 1, ..., t$), i.e. they will be in the rows where h equals $-d_1$, or $-d_2, ...,$ or $-d_t$. The t rows we want to pick out are then

$d_1 - d_1$	$d_2 - d_1$	$\cdot$ $\cdot$ $\cdot$	$d_t - d_1$
$d_1 - d_2$	$d_2 - d_2$	$\cdot$ $\cdot$ $\cdot$	$d_t - d_2$
$d_1 - d_t$	$d_2 - d_t$	$\cdot$ $\cdot$ $\cdot$	$d_t - d_t.$

These rows give the points on the t hyperplanes through x^0. Any other point appears in as many hyperplanes as there are through two points, i.e. $(s^{m-1}-1)/(s-1)$ of them, so that the off-diagonal entries repeat each non-zero value, taken mod $v = (s^{m-1}-1)/(s-1)$, precisely $\lambda = (s^{m-1}-1)/(s-1)$ times: the d_i form a difference set mod v. Hence the Theorem (cf. [90]): If s is the power of a prime, then there exist $t = (s^m-1)/(s-1)$ integers such that each of the non-zero integers mod $v = (s^{m+1}-1)/(s-1)$ occurs as a difference of those integers precisely $\lambda = (s^{m-1}-1)/(s-1)$ times. For $m = 2$ this is called "Singer's theorem" in [7].

Example

$PG(3,2)$, $v = 15$, $t = 7$, $\lambda = 3$.

0	1	12	2	9	13	7
14	0	11	1	8	12	6
3	4	0	5	12	1	10
13	14	10	0	7	11	5
6	7	3	8	0	4	13
2	3	14	4	11	0	9
8	9	5	10	2	6	0

Difference set: 0, 2, 3, 6, 8, 13, 14.

The existence of difference sets for $m = 2$ and s the power of a prime is proved in [25] by using only concepts of finite fields.

If we take a k-flat $(k < m-1)$ with cycle v, then the v k-flats we obtain starting from one of them are not all the k-flats of the space. We can then repeat the procedure, starting from a k-flat not yet considered, and obtain altogether $V(m, k; s)/V(m, 0; s) = w$, say, initial k-flats, (For $k = 1$, this works out as $(s^m-1)/(s^2-1)$.) From these, we can obtain a system of difference sets.

Write the points of the initial k-flats as powers of an element of $GF(s^{m+1})$, as before, thus (again quoting only the exponents)

$$d_{11}, \ldots, d_{1f}$$

$$- \quad - \quad - \quad -$$

$$d_{w1}, \ldots, d_{wf}$$

(where f is the number of points in the k-flat). Then the $wf(f-1)$ differences $d_{ir} - d_{is}$ $(r \neq s)$ contain all non-zero residuals mod v precisely $V(m-2, m-k-1; s)$ times.

58

Example (from [71])

Consider the $PG(4, 2)$. $(2^5 - 1)/(2 - 1) = 31$ and $f = (2^2 - 1)/(2 - 1)$ $= 3$ (the number of points in the space, and on a line, $k = 1$, respectively) are relatively prime. We obtain the difference set $(w = 5)$

0	1	18
0	2	5
0	4	10
0	7	22
0	8	20

Quadrics

A set of points whose co-ordinates satisfy a homogeneous equation of second degree is called a *quadric*. A quadric in $PG(m, s)$ is degenerate if there exists a non-singular transformation of the matrix of coefficients which transforms it into a triangular matrix where all entries in the last row and last column are 0. (Note that a triangular matrix can be transformed into a symmetric matrix of the same size if the characteristic of $GF(s)$ is not 2. If it is 2, then such a transformation may not always be possible.)

We quote here, without proofs, results concerning non-degenerate quadrics, which generalize those about conics in a plane.

It is shown in [68] that a non-degenerate quadric in $PG(2k, s)$ contains $(k-1)$-flats, but no flat of higher dimension, while the highest dimension of a flat in a non-degenerate quadric in a $PG(2k-1, s)$ may be $k-1$, or $k-2$. In the former case the quadric is called *hyperbolic*, or *ruled*, and in the latter *elliptic*, or *unruled*.

The number of different p-flats in a non-degenerate quadric in $PG(m, s)$ is (see [72]):

(i)   for $m = 2k$, $p \leqslant k - 1$ .... $\prod\limits_{r=0}^{p} \dfrac{s^{2(k-p+r)} - 1}{s^{p+1-r} - 1}$

(ii)   for $m = 2k - 1$, $p \leqslant k - 2$, elliptic quadrics

$$\prod\limits_{r=0}^{p} \dfrac{s^{m-2p+2r} + s^{k-p+r-1} - s^{k-p+r} - 1}{s^{p+1-r} - 1}$$

(iii) for $m = 2k - 1$, $p \leqslant k - 1$, hyperbolic quadrics

$$\prod\limits_{r=0}^{p} \dfrac{s^{m-2p+2r} - s^{k-p+r-1} + s^{k-p+r} - 1}{s^{p+1-r} - 1}$$

Ray-Chaudhuri derives these expressions from those in [68] for $p = 0$, i.e. for the enumeration of points. In this latter case we have, respectively,

(i') $\quad (s^m - 1)/(s - 1)$, $\qquad$ (ii') $\quad \dfrac{(s^k + 1)(s^{k-1} - 1)}{s - 1}$

(iii') $\qquad\qquad \dfrac{(s^k - 1)(s^{k-1} + 1)}{s - 1}$

We quote, again from [68], that the number of non-degenerate quadrics in a $PG(m,s)$ is

$$s^{t(t+1)} \prod_{i=1}^{t} (s^{2i+1} - 1) \qquad \text{when } m = 2t, \quad \text{and}$$

$$s^{t(t+1)} \prod_{i=1}^{t-1} (s^{2i+1} - 1) \qquad \text{when } m = 2t - 1.$$

The number of ruled quadrics in a $PG(2t-1,s)$ is

$$\tfrac{1}{2} s^{t^2} (s^t + 1) \prod_{i=1}^{t-1} (s^{2i+1} - 1),$$

and hence that of unruled ones

$$\tfrac{1}{2} s^{t^2} (s^t - 1) \prod_{i=1}^{t-1} (s^{2i+1} - 1).$$

It is also shown in [72] that the number of p-flats in a non-degenerate quadric in $PG(m,s)$ which pass through a given k-flat contained in the quadric equals the number of $(p-k-1)$-flats contained in a non-degenerate quadric of the same type (i.e. elliptic or hyperbolic) in $PG(m-2k-2,s)$. Moreover, to every non-degenerate quadric in $PG(2k,2^t)$ there exists a point, the "nucleus of polarity", outside the quadric, such that every line through this point intersects the quadric in a single point. This is again a generalization of a theorem about a $PG(2,2^t)$.

Further geometric facts about quadrics are given in [73].

Caps

A set of k points such that any t of them are independent, i.e. do not lie in a $(t-2)$-flat, but such that there exists a set of $t+1$ of them that lie in a $(t-1)$-flat, is called a k-cap of type $t-1$. For instance, a conic in a $PG(2,s)$, $s > 2$ is a $(s+1)$-cap of type 2.

A matrix whose $k(\geqslant r)$ rows are the co-ordinates of the points of a k-cap of type $t-1$ in $PG(r-1,s)$ will have rank r (otherwise the points would lie in a flat of lower dimension than $r-1$) and any t rows of it will be independent, while there will be one set of $t+1$ rows which are dependent.

We want to find out how large k can be for a k-cap of type $t-1$ to exist in $PG(m,s)$ and denote the largest possible value of k by $m_t(m,s)$ or, if no confusion can result, simply by m_t. For the derivation we use here an argument contained in [9].

First, let $r+1$ independent points be given in an r-flat, and ask how many "$(r+1)$-stage points" there are in this flat, i.e. points which do not lie in a flat defined by r or less of those given points. We denote the maximal number of $(r+1)$-stage points by P_r.

If $r=1$, then P_1 is the number of points on a line through two given points and not identical with either of them. Hence $P_1 = s-1$.

Assume now that we have proved that in a k-flat there are $(s-1)^k$ different $(k+1)$-stage points, whenever $k < r$. Then, to find the number of $(r+1)$-stage points among the $(s^{r+1}-1)/(s-1)$ points of an r-flat we have to subtract from this number the $r+1$ given points, then for any of the $\binom{r+1}{2}$ lines through any pair of these points the $s-1$ remaining ones, for any of the $\binom{r+1}{3}$ planes through any three of these points the $(s-1)^2$ remaining independent points, and so on. There will remain

$$(s^{r+1}-1)/(s-1) - (r+1) - \binom{r+1}{2}(s-1) - \dots - \binom{r+1}{r}(s-1)^{r-1}$$

points. But $s^{r+1} = \sum_{k=0}^{r+1} \binom{r+1}{k}(s-1)^k$, hence

$$s^{r+1}-1 = \sum_{k=1}^{r} \binom{r+1}{k}(s-1)^k + (s-1)^{r+1},$$

$$(s^{r+1}-1)/(s-1) = \sum_{k=1}^{r}\binom{r+1}{k}(s-1)^{k-1} + (s-1)^r,$$

and therefore $P_r = (s-1)^r$. Thus the latter formula can be proved, by induction, to hold for all integers $r \leqslant m$.

Imagine now that we have found m_t points in a $PG(m,s)$ such that no t lie in a flat of less than $t-1$ dimensions. Each of the $\binom{m_t}{r}$ sets of r points out of the given m_t points determines r-stage points, their number being $(s-1)^{r-1}$.

To continue, we have to distinguish between even and odd values of t. First, let $t = 2h$. Then no $2h$ of the m_t points lie in a space of less than $2h-1$ dimensions, and hence two $(h-1)$-flats, defined by two different sets of h points, have no h-stage points in common. We can therefore group the $(s^{m+1}-1)/(s-1)$ points of the $PG(m,s)$ into 1-stage, 2-stage, ..., h-stage points, and they will all be different.

Hence $\binom{m_t}{1} + \binom{m_t}{2}(s-1) + ... + \binom{m_t}{h}(s-1)^{h-1}$ is not larger than $(s^{m+1}-1)/(s-1)$, i.e. when $t = 2h$, $s^{m+1} \geqslant 1 + \binom{m_t}{1}(s-1) + ... + \binom{m_t}{h}(s-1)^h$.

This argument is still correct if $t = 2h+1$, say, but in this case we can add a further term on the right-hand side. We may not add all the $(h+1)$-stage points of the $\binom{m_t}{h+1}$ sets of $h+1$ points, because they are not all different. But if we choose only those $\binom{m_t-1}{h}$ sets which contain one fixed point, then the other points in them are all different, and we obtain, when $t = 2h+1$, $s^{m+1} \geqslant 1 + \binom{m_t}{1}(s-1) + ... + \binom{m_t}{h}(s-1)^h + \binom{m_t-1}{h}(s-1)^{h+1}$.

These inequalities imply upper limits for m_t.

In some cases the inequality sign can be replaced by an equality sign. For instance, when $t = 2$, then we are looking for the number of points no two of which are the same. These are simply all the points of the $PG(m,s)$. The inequality above reduces to

$$s^{m+1} \geqslant 1 + m_t(s-1), \quad \text{i.e.} \quad m_t \leqslant (s^{m+1}-1)/(s-1)$$

and we see that the equality sign is correct.

Again, if $m = 2$, and $t = 3$, then we ask for the number of points on a plane such that no three are on the same line. We know already that this number is at least that of points on a conic, namely $s+1$, if s is odd. On the other hand, if s is even, then we can add to that number the nucleus of the conic, and then $m_3 \geqslant s+2$. The above inequalities give $s+1$ and $s+2$ respectively, as upper bounds, and we see now that they are reached.

When $s = 2$, then we have $m_3 \leqslant 2^m$. We can show that, once more, the equality sign holds. Consider a $PG(m,s)$ and omit all the points on a hyperplane (we obtain a $EG(m,s)$, see later). There remain two points on each line, and the 2^m points satisfy the requirements. (When $m = 2$, we have $m_3 = 4$, which is also a special case of $s = 2^n$, $m_3 = 2^n + 2$.)

For general m and s, the inequality above produces, for $t = 3$, $m_3 \leqslant 1 + (s^m - 1)/(s - 1)$.

When s is odd, then a sharper upper bound can be found. Let P and Q be two of the m_3 points. There pass $(s^{m-1} - 1)/(s - 1)$ planes through the line PQ, and all points of the $PG(m,s)$ are on one of them. But none of the planes contains more than $s + 2$ out of the m_3 (as we have just seen) and two of these are P and Q, in all planes. Hence

$$m_3 \leqslant 2 + \frac{(s^{m-1} - 1)}{(s - 1)} (s - 1) = 1 + s^{m-1}.$$

If s is odd and $m = 3$, then this gives $m_3 \leqslant 1 + s^2$. But we can prove that for any s, and $m = 3$, we have $m_3 \geqslant 1 + s^2$, so that for odd s equality holds. The proof, again from [9], is as follows.

Consider the quadric surface $ax_0^2 + 2hx_0 x_1 + bx_1^2 = x_2 x_3$ in a $PG(3,s)$, where the left-hand side is irreducible, so that it can only be zero if $x_0 = x_1 = 0$. The plane $x_2 = 0$ meets the surface only in $(0, 0, 0, 1)$, and the plane $x_3 = 0$ in $(0, 0, 1, 0)$. Each of the other $s - 1$ planes of the pencil $x_2 + tx_3 = 0$ meets the surface in a conic with $s + 1$ points, so that there are $(s - 1)(s + 1) + 2 = s^2 + 1$ points on the surface.* No three of these are collinear, because the intersection with any of these planes can be written $ax_0^2 + 2hx_0 x_1 + bx_1^2 + cx_2^2 = 0$, and the left-hand side is irreducible.

In particular, we have $m_3 \geqslant 17$ for $PG(3,4)$. E. Seiden has shown in [85] that in this case $m_3 = 1 + s^2 = 17$ is true (although s is, in this case, even). The proof is elementary, but lengthy, and we shall not reproduce it here.

For even s, we have from above only $m_3(3,s) \leqslant s^2 + s + 2$, but Qvist has proved in [69] that, in fact, $m_3(3,s) = 1 + s^2$ for even s as well (points of an unruled quadric, for example) except for $s = 2$, when 8 points not on a plane form an 8-cap of type 2.

In [100], G. Tallini has proved further relations of which the simplest is $m_t(m,s) = m + 2$ when $\dfrac{s(m+1) - 1}{s + 1} \leqslant t \leqslant m$.

Error-correcting codes

We shall now explain an application of k-caps to a problem of coding.

---

* This agrees with formula (ii′) on page 59.

Consider n-vectors (i.e. vectors with n components) whose components are either 0 or 1. We call the number of unit components the *weight* of the vector, and call the weight of the sum of two vectors v_1 and v_2, when the components are reduced mod 2, or equivalently their difference, their *Hamming distance*, $d(v_1, v_2)$, [36]. This distance is the number of components in which two vectors differ, and it satisfies the triangular inequality $d(v_1, v_2) + d(v_2, v_3) \geqslant d(v_1, v_3)$, because this is true for every component by virtue of the rules of addition mod 2. The weight of a vector is its Hamming distance from the identity element $(0, ..., 0)$.

The 2^n different n-vectors form an Abelian group under addition. Their sub-groups have orders 2^k, $k \leqslant n$. If such a sub-group is given, then we can write down the 2^n different n-vectors in an array of 2^k columns and 2^{n-k} rows, in such a way that the first row contains the elements of the sub-group, and the other rows contain the elements which are the "products" (or rather the vector sums) of the first element of the row, called *coset leader*, and the group element on top of the column, thus:

$$(m = 2^k, \; t = 2^{n-k}) \qquad I, \qquad a_1, ..., a_{m-1}$$
$$c_1, \qquad c_1 a_1, ..., c_1 a_{m-1}$$
$$- \quad - \quad - \quad -$$
$$c_{t-1}, c_{t-1} a_1, ..., c_{t-1} a_{m-1}$$

We may choose any element in a row as the coset leader, without affecting the row, except for permutations of its elements. In particular, assume that the coset leaders have been chosen so that no other element in the row has lower weight. This means that for each coset leader c_i we have $d(c_i, I) \leqslant d(c_i a_j, I)$.

It is easily seen that the distance between two vectors remains the same, if we add the same third vector to the original two. Hence $d(c_i, I) = d(c_i a_k, a_k)$ and $d(c_i a_j, I) = d(c_i a_j c_i a_k, c_i a_k) = d(a_j a_k, c_i a_k) = d(c_i a_k, a_j a_k)$. (For the second equality remember that the group is Abelian, and that $c_i c_i = I$.)

It follows that $d(c_i a_k, a_k) \leqslant d(c_i a_k, a_j a_k)$, which means that any element $c_i a_k$ is at least as near to the group element a_k as it is to any other group element $a_j a_k$.

If we can construct a group of order 2^k of n-vectors with components 0 or 1 and such that all these vectors except the identity $(0, ..., 0)$ have weight $2t + 1$, then an element in a row with coset leader of weight at most t will be nearer to (and not merely not more distant from) the group element on top of the column, than to any other group element.

Hence, if a sequence of n signals, 0 or 1, is transmitted with not more than t errors (which change a 0 into a 1 or a 1 into a 0), then such errors are corrected by assuming that it was the intention to transmit a sequence of signals given by the top element of the column of the array rather than by the signals actually received. Such a system is called a "t-error correcting (n,k) binary group code."

We have here followed the argument given in [91]. The following theorem (from [11]) deals with the existence of error-correcting binary group codes.

If and only if a k-cap of type $2t-1$ exists in a $PG(r-1,2)$, then 2^{k-r} vectors can be found which form a group under addition of the components mod 2, in which all elements except the identity element $(0, ..., 0)$ have at least weight $2t+1$, i.e. at least $2t+1$ components equal to 1.

Let the matrix of k rows and r columns be of rank r and such that any $2t$ rows are independent. This matrix can be transformed, by interchanging rows and/or columns, and by adding or subtracting its columns — all transformations which leave the properties mentioned unaltered — into a matrix $\begin{pmatrix} I_r \\ C \end{pmatrix}$ where C has $k-r$ rows and r columns. Then the matrix $C_0 = (C, I_{k-r})$ has $k-r$ rows and k columns. From its rows we can generate the required group of 2^{k-r} vectors as follows:

Form a vector by adding, mod 2, the components of any $d \leqslant k-r$ rows of C_0. If $d > 2t$, then clearly this vector will have at least weight $2t+1$, from the vectors coming from I_{k-r}. If $d \leqslant 2t$, then we assume that the sum has weight $t_1 < 2t+1$, and derive from this a **contradiction**. Let there be in the portion coming from C precisely c unit components; then $t_1 = c+d$. Corresponding to each of the c positions, we can find a vector in I_r with unity in that position only. Then these c rows from I_r, and d rows of C, constitute a set of $c+d$ vectors which are dependent. But $c+d = t_1 < 2t+1$ contradicts the assumption that in the original matrix any $2t$ vectors were independent.

Together with the identity element $(0, ..., 0)$ we obtain by adding $1, 2, ..., k-2$ vectors, altogether $1 + \begin{pmatrix} k-r \\ 1 \end{pmatrix} + ... + \begin{pmatrix} k-r \\ k-r \end{pmatrix} = 2^{k-r}$ vectors which are all different, because their last $k-r$ components are. Hence they form a group with the required property. The proof of the converse proceeds by reversing the steps of this argument.

In this way we can construct a t-error correcting $(r, k-r)$ binary group code.

Euclidean geometries

The spaces, or geometries, so far considered have been called projective because of their analogy to the customary continuous projective geometry. In the same way, as we relate a continuous Euclidean space to a continuous projective one, we can define a finite Euclidean geometry, denoted by $EG(m,s)$, as the aggregate of flats which remain when a hyperplane with all its flats is removed from a $PG(m,s)$. Those flats which were removed are called *flats at infinity*. Those remaining flats which intersect in a flat at infinity are called *parallel*.

It is convenient to consider the excluded hyperplane as that whose equation is $x_0 = 0$. Then, since the sets $(x_0, ..., x_m)$ and $(cx_0, ..., cx_m)$ with $c \neq 0$ were considered to denote the same point, we may fix x_0 for all points in the $EG(m,s)$ at 1, and consider only the remaining coordinates as those of a point in $EG(m,s)$.

It is readily seen that flats of a $PG(m,s)$ not at infinity are also equivalent to finite Euclidean geometries. Their numbers are easily obtained. For instance, since there are $s^m + ... + s + 1$ points in a $PG(m,s)$, and the $s^{m-1} + ... + s + 1$ points of a hyperplane were removed, there remain s^m points in the $EG(m,s)$. Generally, there are $V(m,k,s) - V(m-1,k;s)$ remaining k-flats. This is also the number of k-flats in an m-flat in a geometry of higher dimension. The number of t-flats through a k-flat is the same as in a $PG(m,s)$

A k-flat within a $EG(m,s)$ contains all those s^k points which satisfy $m-k$ equations of the form $a_{0j} + a_{1j}x_1 + ... + a_{mj}x_m = 0$ $(j = 1, ..., m-k)$. In particular, a hyperplane is defined by $a_0 + a_1x_1 + ... + a_mx_m = 0$. If we keep $a_1, ..., a_m$ constant and give to a_0 all the s values of the $GF(s)$, then we obtain a pencil of parallel hyperplanes. All the hyperplanes of the pencil pass through a $(m-2)$-flat at infinity, defined, in terms of the original $PG(m,s)$, by $x_0 = 0$, $a_0x_0 + a_1x_1 + ... + a_mx_m = 0$, with the given $a_1, ..., a_m$. No two hyperplanes of such a pencil have any point in common, and through every point of the $EG(m,s)$ there passes precisely one hyperplane of the pencil.

If we consider a pencil of hyperplanes in a Euclidean space as determined by $a_1, ..., a_m$, then $ca_1, ..., ca_m$ $(c \neq 0)$ determine the same pencil, though the individual hyperplanes need not remain the same. Hence there are $(s^m - 1)/(s - 1)$ different parallel pencils of hyperplanes in the $EG(m,s)$.

More generally, there are $(s^{k+1} - 1)/(s - 1)$ different $(m-k)$-flats through a $(m-k-1)$-flat at infinity, and $(s^k - 1)/(s - 1)$ of these are entirely at infinity, while s^k belong to the $EG(m,s)$.

The s^m points of a $EG(m,s)$ can be represented by $v = s^m - 1$ powers of x, and 0. A k-flat not through 0 contains all the points $\sum_{i=0}^{k} a_i x^{b_i}$, where the a_i are elements of $GF(s)$, $\sum_i a_i = 1$, and the x^{b_i} are $k+1$ independent points. A k-flat through 0 contains all points $\sum_{i=1}^{k} a_i x^{b_i}$, without the restriction on the sum of the a_i.

Consider a flat of the former type. It follows from an argument similar to that used for projective spaces that the cycle of such a flat is $s^m - 1$, since $s^m - 1$ and s^k are always relatively prime.

The s elements of $GF(s)$ can be written $0, x^0, x^w, \ldots, x^{(s-2)w}$, where $w = (s^m - 1)/(s-1)$, and if x^c is on a k-flat through 0, then so is x^{c+iw}.

A pair of points x^{c_1} and x^{c_2} will occur in the same flat through 0 if and only if $c_1 - c_2 = 0 \pmod{w}$. They will occur together

$$V(m - 2, m - k - 1; s)$$

times in the same k-flat.

If $c_1 - c_2 \neq 0 \pmod{w}$, then the three points $0, x^{c_1}$ and x^{c_2} occur together $V(m-3, m-k-1; s)$ times in the same k-flat, so that the two latter points occur together in the same k-flat, not containing 0,

$$[V(m-2, m-k-1; s) \; - \; V(m-3, m-k-1; s)] \quad \text{times.}$$

The difference sets constructed from initial k-flats will produce all differences and $v = s^m - 1$ the same number of times, except those divisible by w, which are not produced at all.

Example

$EG(3,3)$, $v = 26$, $w = 13$, $k = 1$. The initial lines are

  0 1 22   0 2 8   0 3 14   0 7 17

The difference 13 does not appear, all others appear once.

This example and the statements preceding them are taken from [71], where the construction of yet more general difference sets is also dealt with.

In particular, take lines in a plane, so that $v = s^2 - 1$, and $w = s + 1$. Then we can find s integers such that all positive integers less than $s^2 - 1$ and not divisible by $s+1$ appear once among their differences, while the others do not appear at all. This is the "affine analogue to Singer's theorem" (see [7]).

From the paper just referred to, we quote a few difference sets:

s	s^2-1	$s+1$	Difference set	Differences
3	8	4	1,6,7	5,6,1,3,2,7
4	15	5	1,3,4,12	2,3,11,
				1,9,8,
				13,12,4,
				14,6,7
5	24	6	1,3,16,17,20	(See Exercises)
7	48	8	1,2,5,11,31,36,38	(See Exercises)

EXERCISES

(1) Describe the structure of a $PG(4,3)$.

(2) Determine the 3-flats through the plane given by

$$(1,0,0,0,0), \quad (0,0,1,0,0), \quad \text{and} \quad (0,0,0,0,1)$$

in a $PG(4,3)$, and state which of them are independent.

(3) Let $(a_{ij}) = \begin{pmatrix} 0 & -1 & 1 \\ 1 & 0 & -1 \\ -1 & 1 & 0 \end{pmatrix}$.

Show the correspondence between points and lines defined by

$$\sum_i \sum_j a_{ij} x_i y_j = 0 \text{ in a } PG(2,2).$$

(4) Answer Exercise (3) for the following matrix (a_{ij}):

$$\begin{pmatrix} 0 & -1 & 2 \\ 1 & 0 & 1 \\ -2 & -1 & 0 \end{pmatrix} \text{ in a } PG(2,3).$$

(5) Construct an 8-cap of type 2 in a $PG(3,2)$.

(6) Find the differences for $s=5$ and $s=7$ from the difference set at the end of Chapter III.
Mention the differences which do not appear.

CHAPTER IV

CONFIGURATIONS

The designs we wish to consider consist of two sets of objects, with an incidence relation between objects of different sets. For instance, the objects may be points and lines, with a given point lying or not lying on a given line, and a given line containing or not containing a given point.

We shall call the two types of objects *varieties* and *blocks*. These expressions are taken from agricultural field trials, where these ideas were first applied to statistical experiments. The number of varieties will, as a rule, be denoted by v, and that of the blocks by b.

We shall always assume that no variety appears more than once in the same block, and that no two blocks are identical. Also, to exclude trivial cases, we assume $v > 1$ and $b > 1$.

A special design is that where there are k varieties in each block and every variety appears in r blocks. We call this design a *configuration*, and denote it by (b_k, v_r). Clearly, $bk = vr$.

Such a configuration is a special case of a tactical configuration of rank n, defined by E.H. Moore in [57] as a collection of n sets of objects, with an incidence relation such that every object of a set a is incident with the same number of objects of set b, for all $a \neq b$.

All the flats of a $PG(m, s)$, or of a $EG(m, s)$ form a tactical configuration of rank m. When $g < h$, then there are $V(h, g; s)$ g-flats in any h-flat of a $PG(m, s)$, and there are $V(m-g-1, m-h-1; s) = V(m-g-1, h-g-1, s)$ different h-flats through any given g-flat, as we know.

In this terminology, a configuration (b_k, v_r) is a tactical configuration of rank 2. We shall only be concerned with this rank, and we shall therefore use only the shorter term, configuration.

If $b = v$, and hence $r = k$, then we call the configuration *symmetric*, and denote it by (v_r). The (v_2) are the v-sided polygons. When $g + h = m - 1$, then $V(m, g; s) = V(m, h; s)$, and the g-flats and h-flats $(g < h)$ form a symmetric configuration. In particular, the points and lines of a $PG(2, s)$ form a configuration with $v = b = 1 + s + s^2$, and $r = k = 1 + s$. In a $PG(2,2)$, we have thus a (7_3), and this is the only configuration of this type (apart from isomorphisms).

A (v_3) exists only for $v \geqslant 7$. There is also essentially only one (8_3), which can be represented as in Fig. 9a or 9b. (If we wish to draw a unit circle so that the upper and lower lines are tangents, then, placing the origin of a system of co-ordinates into the centre of the circle and taking the line through 3 and 7 as the x-axis, the abscissa x_1 of point 1 and the abscissa x_3 of point 3 must satisfy the equation $3(x_1 - x_3) = 2(x_1 x_3 - 1)$. Fig. 9b is the case for $x_1 = 0$.)

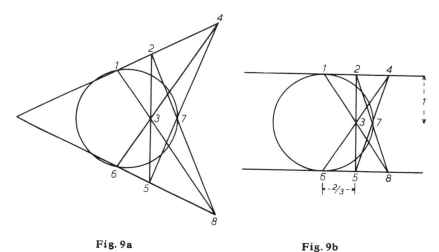

Fig. 9a   **Fig. 9b**

There are three essentially different (9_3) (see [45]). One of them has been represented in Fig. 4, and the other two may be drawn as in Fig. 10a and 10b.

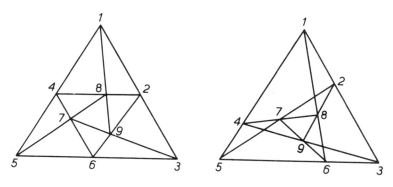

Fig. 10a   **Fig. 10b**

(If the lines are to be straight, then the points must be judiciously chosen. For instance, if in Fig. 10a $\overline{135}$ and $\overline{246}$ are equilateral triangles, then the ratio $\overline{28}$ to $\overline{48}$ is $(3-\sqrt{5})/(\sqrt{5}-1)$).

One of the ten existing $(10_3)^*$ is that illustrating the theorem of Desargues (in Fig. 8, the points $ABCabcPQRS$). We obtain the same diagram by taking five points in space, no four of them in the same plane, with all their connecting lines, and all the planes through any three of them, and intersecting this by a plane that does not pass through any of the intersections of three of these planes.

We mention, without any further detail, that there exist five (8_4) and twenty-six (9_4) (cf. [42], p. 118. This book contains a masterly survey of configurations, and in particular a description of Reye's (12_6) and of his $(16_3, 12_4)$ as well).

If we omit, from a $PG(2, s)$, all points on two lines and all lines through their common points, and also all lines through one of the other points on the original two lines, then we obtain an $(s(s-1)_{s-1})$.

If a design has the property that any $\mu(\geqslant 2)$ varieties appear together λ_μ times in the same block, then we call it μ-wise (if $\mu = 2$, pairwise) balanced. A μ-wise balanced configuration has also been called "complete" and denoted by $C(\lambda_\mu, \mu, k, v)$. It is not necessary to indicate r and b explicitly, because $b = rv/k$, and r can be derived from λ_μ and μ as follows.

Any $\mu - 1$ appear also the same number of times, say $\lambda_{\mu-1}$ times, together in the same block. In each of these blocks there are $k - \mu + 1$ varieties other than those $\mu - 1$ chosen, and these are therefore $\lambda_{\mu-1}(k - \mu + 1)$ in number, but, of course, not all different. On the other hand, each of the $v - \mu + 1$ varieties which were not chosen appears in these blocks λ_μ times, since this is the number any of them appears together with the $\mu - 1$ chosen elements, and this must happen within the $\lambda_{\mu-1}$ blocks which we consider. Hence

$$\lambda_{\mu-1}(k - \mu + 1) = \lambda_\mu(v - \mu + 1).$$

It follows that $\lambda_{\mu-1} = \lambda_\mu(v-\mu+1)/(k-\mu+1)$, and this number is independent of the particular set of $\mu - 1$ varieties from which we started.

Continuing, we obtain $\lambda_{\mu-2}(k-\mu+2) = \lambda_{\mu-1}(v-\mu+2)$ and so on, and eventually $\lambda_1(k-1) = \lambda_2(v-1)$. By back-substitution, we obtain

$$\lambda_\mu = \lambda_\tau \binom{k-\tau}{\mu-\tau} \bigg/ \binom{v-\tau}{\mu-\tau}, \quad \text{for } \tau = 1, 2, \dots, \mu-1.$$

---

* They are all illustrated in [45].

It follows that in a $C(\lambda_\mu, \mu, k, v)$ the numbers

$$\lambda_\mu \binom{v - \tau}{\mu - \tau} \bigg/ \binom{k - \tau}{\mu - \tau}$$

are integers for $\tau = 1, ..., \mu - 1$.

λ_1 is the same as r, and from now on we write λ instead of λ_2. We note, in particular, that $b = vr/k = \lambda_\mu \binom{v}{\mu} \bigg/ \binom{k}{\mu}$, which must also be an integer. For consistency, we can write $\lambda_0 = b$.

Our proof implies that if a design is balanced, and if all blocks contain the same number of varieties, then it is a configuration, and all varieties are equally often repeated.

If all combinations of k varieties out of v appear together in one block, then

$$b = \binom{v}{k}, \quad r = \binom{v - 1}{k - 1}, \quad \lambda = \binom{v - 2}{k - 2}, \quad ...., \quad \lambda_k = 1,$$

and this can also be derived from first principles.

A $C(\lambda_\mu, \mu, k, v)$ is also a $C(\lambda_\tau, \tau, k, v)$ for $0 < \tau < \mu$. In particular, when $\mu \geqslant 2$, then it is a $C(\lambda, 2, k, v)$, a so-called *balanced incomplete block design* (b.i.b.d.) (called "incomplete randomized blocks" by F. Yates in [106]).

Example

| 0 1 2 | 0 1 3 | 0 2 4 | 0 3 5 | 0 4 5 |
| 1 2 5 | 1 3 4 | 1 4 5 | 2 3 4 | 2 3 5 |

This design appeared in [107] and is perhaps the first b.i.b.d. to have been published in the literature.

The points of a $PG(m, s)$, or of a $EG(m, s)$ considered as varieties, and its t-flats considered as blocks are such b.i.b.d's, with the following parameters:

$PG(m, s)$		$EG(m, s)$
$V(m, 0;\ s)$	v	s^m
$V(m, t;\ s)$	b	$V(m-1, t-1;\ s)s^{m-t}$
$V(m-1, t-1;\ s)$	r	$V(m-1, t-1;\ s)$
$V(m-2, t-2;\ s)$ if $t \neq 1$, or	k	$V(m-2, t-2;\ s)$ if $t \neq 1$, or
1, if $t = 1$		1, if $t = 1$.

When $m = 2$, then b.i.b.d's can be constructed from non-Desarguesian finite planes as well.

A symmetric b.i.b.d. is also called a (v, k, λ) *configuration*.

Construction

Given a system of difference sets $\bmod\, v$ (the power of a prime)

$$(d_{i1}, \ldots, d_{ik}) \qquad (i = 1, \ldots, t)$$

we can take the elements as varieties, and the blocks

$$(d_{i1} + d, \ldots, d_{ik} + d)$$

where d runs through all the elements of $GF(v)$. We then obtain a b.i.b.d. with $b = tv$, $r = tk$.

Proof: Every element, say x, appears as often as $d_{ij} + d = x$ holds. Since to every d_{ij} there is a d satisfying this equation, this will happen tk times. Each pair of elements, say x and y, appears as often in the same block as $d_{ij} + d = x$, and also $d_{ik} + d = y$ holds for some j and k, i.e. as often as $d_{ij} - d_{ik} = x - y$, with given x and y. This difference appears as often as any other. (Cf. [6]).

Examples

(1) From Sprott's system of difference sets (see [93]) mentioned on page 23 we obtain a b.i.b.d. with $v = 6t + 1$, $b = t(6t + 1)$, $k = 3$, $r = 3t$, provided $6t + 1$ is a prime power.

(2) If $v = 6t + 1$ is a prime, then a b.i.b.d. with $b = t(6t + 1)$, $k = 3$, $r = 3t$ is obtained from the system of difference sets

$$(0, x^a, x^{a+t})$$

where x is a prime element of $GF(v)$, and $a = 0, \ldots, t - 1$. For the proof, see [15] p. 427. If $t = 1$, then the b.i.b.d. obtained is symmetric.

If $d_1, \ldots, d_k$ is a difference set $\bmod\, v$, then so is $-d_1, \ldots, -d_k$, and the b.i.b.d. arising from the latter is "dual" (see page 80) to that arising from the former.

Chowla and Ryser (cf. [18]) have pointed out that there exist symmetric b.i.b.d's not derivable from difference sets. For instance, a Hadamard matrix of order 56 exists, as we have seen, and hence a b.i.b.d. with $v = b = 55$, $r = k = 27$, $\lambda = 13$ can be constructed, as is shown in [103], on page 14. However, no difference set exists of 27 numbers mod 55.

In [6] Bose introduced a construction which we shall illustrate by an example.

The varieties are given by an ordered pair of numbers (x, y), where x equals a residual mod $2t+1$, and $y = 1, 2$, or 3. We have then $v = 6t+3$ different varieties.

Consider the basic blocks of 3 varieties each

$$
\begin{array}{ccc}
(i, 1) & (-i, 1) & (0, 2) \\
(i, 2) & (-i, 2) & (0, 3) \\
(i, 3) & (-i, 3) & (0, 1)
\end{array}
\quad \text{with } i = 1, 2, \ldots, t
$$

and $\quad (0, 1) \quad (0, 2) \quad (0, 3)$.

These are altogether $3t+1$ blocks. If we add to i the values $m = 1, 2, \ldots, 2t$, then we obtain altogether $b = (2t+1)(3t+1)$ blocks, and they form a b.i.b.d. with $k = 3$, $r = 3t+1$, $\lambda = 1$, i.e. a $C(1, 2, 3, 6t+3)$, as will now be shown.

We have to show first that all varieties appear r times, and then that all pairs of varieties appear once in the same block.

The variety $(0, 1)$ appears once in the last basic block, t times in the basic blocks $(i, 3)(-i, 3)(0, 1)$, and in the derived blocks whenever $i+m = 2t+1$, or whenever $-i+m = 2t+1$, that is altogether $3t+1$ times. The same demonstration applies to $(0, 2)$ and to $(0, 3)$. The proof for any other variety is similar, but follows more simply from the fact that they all appear equally often.

The proof of $\lambda = 1$ is equally easy and is here omitted.

A similar principle, using a further element which is not altered by addition, is also described in [6].

B.i.b.d's are treated in great detail in [103].

Steiner triple systems

The special case of a $C(1, 2, 3, v)$ is called a "Steiner triple system". For instance, the $PG(m, 2)$ and the $EG(m, 3)$ are such systems. Jacob Steiner was led to considering these triple systems in [98] from a study of the bi-tangents of a quartic plane curve.

Example

$C(1, 2, 3, 9)$

$$
\begin{array}{cccc}
123 & 145 & 167 & 189 \\
468 & 269 & 258 & 247 \\
579 & 378 & 349 & 356 \, .
\end{array}
$$

These same triples are also obtained as the points of inflection of a plane cubic which lie on the same straight line.

We may ask for which values of v such triple systems exist. From the relations $bk = vr$ and $\lambda(v-1) = r(k-1)$ we have $3b = vr$ and $v-1 = 2r$, hence

$$v = 2r+1 \quad \text{and} \quad b = r(2r+1)/3 = v(v-1)/6.$$

The latter must be an integer, so that either r or $2r+1$ must be a multiple of 3. In the former case

$$r = 3t, \quad b = t(6t+1), \quad v = 6t+1$$

and in the latter

$$r = 3t+1, \quad b = (2t+1)(3t+1), \quad v = 6t+3.$$

(It is easily seen that the values $2^{m+1}-1$, i.e. the numbers of points in a $PG(m,2)$ for varying m, are alternately of the form $6t+1$ and $6t+3$, while the number of points in a $EG(m,3)$, namely 3^m, is an odd multiple of 3, and hence always of the form $6t+3$.)

It is remarkable that for every v of the form $6t+1$ or $6t+3$ a configuration $C(1,2,3,v)$ exists.

We already know from Bose's construction just described that a $C(1,2,3,6t+3)$ exists for all t. We have also seen that a $C(1,2,3,6t+1)$ exists for all t, provided $6t+1$ is a prime number. We must still show that it exists for all other values of $6t+1$ as well (e.g. for $6t+1 = 55$).

To begin with, we remark that from a system with v varieties $x_1,...,x_v$ we can construct one with $2v+1$ varieties. Denote $v+1$ additional varieties by $x_0', x_1', ..., x_v'$. From a given triple $x_i x_j x_k$ form three more, namely $x_i' x_j' x_k$, $x_i' x_j x_k'$, and $x_i x_j' x_k'$. Moreover, form the triples $x_0' x_u x_u'$ for $u = 1, 2, ..., v$. From the original $b = v(v-1)/6$ triples we obtain thus $4b+v = (2v+1)2v/6$ triples, and it is seen that they form a Steiner triple system.

By this procedure we can form a system with $v = 6(2t+1)+1$ from that with $v = 6t+3$.

M. Reiss has shown in [74] how to construct a system with $2v-5$ varieties from one with v varieties. We refer the reader for the proof to that paper, or to [61]. By this method we obtain a system with $6(2t)+1$ from one with $6t+3$. Hence any system with $6t+1$ varieties can be constructed.

Other methods of constructing Steiner triple systems are given in, amongst other books, [15], pp. 425 ff.

E.H. Moore has shown in [56] that at least two non-isomorphic systems exist for all $v > 9$; two are known for $v = 13$, and 80 were found by automatic computing methods (cf. [31]). There exists only one for 3, 7, and 9. That for 3 is trivial, that for 7 is the (7_3), and that for 9 is the resolvable configuration $C(1,2,3,9)$ given on page 73, above. A system for $v = 19$ and one for $v = 21$ (due to R.A. Fisher) is exhibited in [10].

A thorough investigation of the structure of Steiner triple systems and of their isomorphisms is contained in [30].

More general $C(1,\mu,k,v)$ were considered by Moore in [57]. Sprott lists various properties of $C(1,\mu,k,v)$ in [94]. It has been shown in [38] that it is not only necessary but also sufficient for the existence of a $C(\lambda_3,3,4,v)$ that $\lambda_3 \binom{v-\tau}{3-\tau} \Big/ \binom{4-\tau}{3-\tau}$ be integer for $\tau = 0,1,2$ for any value of λ_3, and for $\lambda_3 = 1$ this has already been shown in [37]. In the latter case this means that v is of the form $2^t(3^s+1)(t = 0, 1, \ldots; s = 1, 2, \ldots)$, i.e. of the form $6u+2$, or $6u+4$.

Example

If we take the 8 points of the $EG(3,2)$ as varieties and the 14 planes as blocks, then we obtain the following $C(1,3,4,8)$:

| 1234 | 1256 | 1278 | 1357 | 1368 | 1458 | 1467 |
| 5678 | 3478 | 3456 | 2468 | 2457 | 2367 | 2358 |

This system was investigated by Moore in [58].

Derived designs

Imagine that we omit one of the varieties, say x, from all the blocks of a $C(\lambda_\mu,\mu,k,v)$ and consider the two sets of blocks which we obtain, the first, C_1 say, containing those blocks of k varieties which remain (those which did not contain x in the first place), and the second, C_2 say, containing blocks of $k-1$ varieties. C_1 as well as C_2 are again complete configurations, this time with $v-1$ varieties.

Denote the parameters of C_1 by v', b', r', k' and $\lambda'_\tau(\tau = 2, 3, \ldots, \mu-1)$. Then $v' = v-1$, $k' = k$. Those blocks which did not contain x originally number $b' = b-r$ and λ_τ changes into $\lambda'_\tau = \lambda_\tau - \lambda_{\tau+1}$. The second term on the right-hand side is the number of τ-tuples in the original blocks which appeared together with x. In particular, $r' = r - \lambda$.

Denote the parameters of C_2 by $v'', b'', r'', k'', \lambda_\tau''$. Then $v'' = v-1$, $b'' = r$, $k'' = k-1$, $\lambda_\tau'' = \lambda_{\tau+1}$.

If a configuration is given, then we construct its "complementary" design by replacing each block by a block containing those varieties which are missing from the former. The number of varieties and blocks remains unaltered, say v and b. The number of varieties in a new block is $v-k$, where k is that in the original blocks. To find r, imagine two designs written side by side, with corresponding blocks in the same row. This gives a pattern of b rows and v columns, in which each variety is b times repeated. It was r times repeated in the complementary one. Also, each pair of varieties appears in each row of the combined pattern, i.e. altogether b times in the same row. To compute how often a particular pair appears in the same block of the new design, we have to reduce b by the number of appearances of either variety in the original design, i.e. by r for either of them. But if the two varieties appeared together ν times in the same original block, then ν blocks were now counted twice. Hence in the complementary design such a pair appears $b-2r+\nu$ times in the same block. If the original design was balanced, then $\nu = \lambda$ for any pair of varieties. Similar formulae can be derived for any n-tuple of varieties.

Representations

There are various ways in which a design can be represented. The most obvious representation consists of listing, for the several blocks, the varieties which are contained in them. It is understood that a permutation of the varieties within the same block, and also a permutation in the order of the blocks, does not change the configuration.

If we use this representation for a symmetric configuration, then we can write the varieties within each block in such a way that each variety appears only once as the first written in the block, once as the second, and so on. Such arrangements are called *Youden squares*, after the author of [108] who introduced them. It was conjectured by R.A. Fisher and F. Yates that this is always possible, and for its proof we refer to the following theorem of P. Hall:

Given sets $S_1, \dots, S_m$ of elements from $\{1, 2, \dots, n\}$ it is possible to find m different elements, one from each set, if and only if any $p \, (\leqslant m)$ of the sets contain between them at least p elements.

The necessity of the condition is obvious; its sufficiency was proved by P. Hall in [35], and it is also proved in [102] pp. 64-65, as a consequence of the Duality theorem in linear programming. We call the chosen elements, if they exist, a *system of distinct representatives* (SDR).

Any p blocks of k varieties each of a symmetric configuration contain at least p different varieties between them, because they contain altogether $p.k$, and although these are not all different, none of them can appear more than $r = k$ times. Hence the number of different varieties is at least $pk/k = p$. We can therefore choose a SDR from the b blocks and consider them as being the first varieties in their respective blocks. Each variety will appear another $k-1$ times in the remainders of the blocks, and we repeat the argument to obtain the second, ..., kth variety in each block.

The possibility of a Youden square arrangement can also be proved by applying a theorem of the theory of graphs. A graph in which the points can be grouped into two sets, each arc joining a point of one set to a point of the other, and such that the same number of arcs meet in every point, is called an *even regular graph*. The theorem we are referring to states that every such graph can be obtained as a superposition of even regular graphs with just one line at each point.

A symmetric configuration can be represented as an even regular graph with $k = r$ lines at each point, as follows:

Draw a point for each variety, and also one for each block. Then connect each point representing a variety to the point representing a block in which that variety occurs. In a (b_k, v_r) we shall then have r lines at each point of the first class and k lines at each of the second. When $k = r$, we obtain an even regular graph.

Its k subgraphs can be considered as connecting points for blocks severally to the first, second, ..., variety in them, and the result follows.

A constructive proof of the theorem is contained in [92]. In [39] the authors point out that if $b = m.v$, then each variety can be made to occur m times in the same position by splitting the $r = mk$ replications into k replications of mv pseudo-varieties. Moreover, if $r = mk+1$ (or, alternatively, if $r = mk-1$, $m \geqslant 2$), then it is possible to group the varieties into k groups, so that, after suitable rearrangements within the blocks, there appear in each position either m or $m+1$ varieties (or, in the alternative case, either m or $m-1$ varieties).

We have given graphical representations of configurations in Fig. 1, 9a, 9b and 10a, 10b. Varieties and blocks are given by points and lines respectively in a continuous Euclidean plane. If we want the lines to be straight and infinite, and also such that any two intersect only in a point of the configuration, then only (i) a (3_2), (ii) a $(b_1, 1_b)$ and (iii) a $(1_k, k_1)$ can be drawn, i.e. (i) a triangle, (ii) b lines intersecting in one point, and (iii) a line with k points on it. This is proved in [50] as follows:

Let a_0 be the number of points, a_1 that of line segments, and a_2 that of faces (polygons not intersected by lines). A segment extending to infinity is counted as the same as that on the other end of the line, and an analogous remark applies to faces.

The b lines intersect in $\binom{b}{2}$ points, but they are not all different, since each point lies on $\binom{r}{2}$ different pairs of lines. Hence $a_0 = \dfrac{b(b-1)}{r(r-1)}$.

In each point there intersect $2r$ segments, and each segment is counted twice – once for each of its end-points – so that $a_1 = a_0 \cdot r$.

We have, also, $a_0 + a_2 = a_1 + 1$. This is true for two lines $(a_0 = 1, a_1 = 2, a_2 = 2)$ and an additional line does not alter this relation.

Hence $a_2 = a_1 + 1 \dfrac{a_1}{r} = a_1(r-1)/r + 1$. If $r = 1$, then $a_2 = 1$, and $a_1 = a_0$, i.e. one line with a_1 points on it (case iii above).

If $r = 2$, then we have a (3_2) (case i) or a $(2_1, 1_2)$. If $r > 2$, then $3a_2 = \dfrac{3(r-1)}{r} a_1 + 3 > 2a_1$. But unless all lines pass through the same point (case ii), $2a_1$ is the number of all sides of all faces, which must at least be $3a_2$. We thus reach a contradiction. No other representation of configurations of the type described is possible.

Other configurations can be represented in space. Thus the vertices and edges of a tetrahedron form a $(6_2, 4_3)$. If we project this into a plane, then we obtain, for instance, a square with its diagonals, but their intersection is not a point of the configuration. Its dual, a $(4_3, 6_2)$, is a quadrilateral with its six points of intersection.

A restriction on the representation of balanced configurations by diagrams is also implied in the following theorem proved in [19] (as a corollary).

Theorem 4.1   If any two points are to be connected by straight lines, then there must be at least as many lines as points, unless all points are on the same line; the numbers of points and lines are only equal if all points but one are on the same line.

Proof: The authors prove first the following lemma. If there are n points in a (continuous) projective plane, such that any line through two points contains a third point, then all points lie on one and the same straight line. (This was conjectured by J. Sylvester, and the proof here quoted is due to Gallai. For further historical detail, see [19].)

Assume that the points are not all on the same straight line. Project one point, P_0, to infinity and connect it to all the others, thus obtaining a parallel pencil. Let the line L through $P_i P_j P_k$, in this order, form the smallest positive angle with the parallels. But the line $P_j P_0$ contains at least one other finite point, say P_r, and either $P_i P_r$ or $P_k P_r$ forms a smaller angle with the parallels than does L. This contradiction establishes the lemma.

It follows, for instance, that a (7_3), which is balanced with $\lambda = 1$, cannot be represented by straight lines only. The necessity for a curve in a diagram for (8_3) is mentioned in [42], p. 90.

We use the lemma to establish Theorem 4.1. It is true for three points, in a trivial way. Assume that it is true for any $n-1$ points. Consider now n points $P_1, \ldots, P_n$, not all on the same straight line. Let the straight line through $P_1 P_2$ be that with no other of the n points on it (it exists by virtue of the lemma). If $P_2, \ldots, P_n$ are all on the same straight line, then $P_1 P_i$ $(i = 2, \ldots, n)$ and $P_2 P_3 \ldots P_n$ are precisely n straight lines, and any two points are connected by one of them.

On the other hand, if $P_2, \ldots, P_n$ do not lie on the same straight line, then they determine at least $n-1$ straight lines, and $P_1 P_2$ is not one of them. Thus we again have at least n straight lines. In fact, we now have altogether more than n lines, i.e. not only $P_1 P_2$ is added, but also $P_1 P_3$, $P_1 P_4$, ..., unless $P_3, \ldots, P_n$ are all on the same line, and P_1 lies on it as well. This establishes the complete theorem.

Incidence matrix

A very powerful way of representing a design is the incidence matrix. It is defined as a v by b matrix of 1 and 0, where the rows correspond to the v varieties, the columns to the b blocks, and the entry in the ith row and jth column is 1 if the ith variety appears in the jth row, and is otherwise 0.

The inner product of any column by itself is the number of varieties in the block corresponding to that column (i.e. k in a configuration). The inner product of two different columns is the number of varieties common to the two blocks to which the columns refer, say μ_{st} for blocks s and t. Thus, if we call the incidence matrix A, we have $A'A = (\mu_{st}).(\mu_{ss} = k)$. This is a symmetric matrix of order b. Roy and Laha call it, in [76], the *block structure matrix*.

The inner product of any row of the incidence matrix by itself is the number of repetitions of the variety to which the row refers, i.e. r in a configuration. The inner product of two different rows is the number of times the two varieties to which the rows refer appear together in the same block, say λ_{uv} for varieties u and v. Thus $AA' = (\lambda_{uv})(\lambda_{uu} = r)$. This is a symmetric matrix of order v, which Roy and Laha call the *treatment structure matrix* (treatment being a synonym for variety).

If all λ_{uv} have the same value, say λ, then the design is (pair-wise) balanced. If all μ_{st} have the same value, say μ, then the design is called of *linked block type* (see [109]).

Two designs with incidence matrices A and A' respectively are called *dual*. If we change 1 into 0 and 0 into 1, then we obtain the incidence matrix of a complementary design.

For a configuration, the total of the elements of the wth column of AA' is $r + \sum_{\substack{u=1 \\ u \neq w}}^{v} \lambda_{uw}$. The second term gives the number of all appearances of variety w together with any of the other $k-1$ varieties in the same block. This number is $r(k-1)$, hence the total of each column is $r + r(k-1) = rk$, independently of w. All columns (and rows) have the same total. For a balanced configuration, we have $r + \lambda(v-1) = rk$.

To compute the determinant $|AA'|$ when A is the incidence matrix of a balanced incomplete block design, i.e. to compute

$$\begin{vmatrix} r & \lambda & \lambda & \lambda \\ \lambda & r & \lambda & \lambda \\ & & \cdot & \\ & & & \cdot \\ \lambda & \lambda & \lambda & r \end{vmatrix}$$

subtract the first column from the others, to obtain

$$\begin{vmatrix} r & \lambda-r & & \lambda-r \\ \lambda & r-\lambda & & 0 \\ & & \cdots & \\ \lambda & 0 & & r-\lambda \end{vmatrix}$$

and then add all the other rows to the first. The result is

$$|AA'| = \begin{vmatrix} rk & 0 & 0 \\ \lambda & r-\lambda & 0 \\ & \cdots & \\ \lambda & 0 & r-\lambda \end{vmatrix} = rk(r-\lambda)^{v-1}.$$

Proceeding in the same way, we find that the latent roots of AA' are rk (once) and $r-\lambda$, $((v-1)$-fold). The latent vector $(x_1, ..., x_v)$ corresponding to the latent root rk is found by solving the system:

$$(r-rk)x_1 + \lambda x_2 + \ \ldots \ + \lambda x_v = 0$$
$$\lambda x_1 + (r-rk)x_2 + ... + \lambda x_v = 0$$
$$- - - - - - - -$$
$$\lambda x_1 + \lambda x_2 + ... + (r-rk)x_v = 0.$$

Because of $r-rk = \lambda(1-v)$, this is satisfied by $x_1 = x_2 = ... = x_v = 1$. An analogous statement refers to the determinant and the latent roots of $A'A$, when A is the incidence matrix of a linked block type design with equal numbers in all blocks.

Structure

The concept of the incidence matrix will now be used to demonstrate structural features of general designs, and of configurations.

Theorem 4.2   In a balanced design the number of blocks, b say, must be at least that of the varieties, say v.

The following proof is taken from [43], where actually the dual statement is proved, namely that in any linked-type design the number of varieties is at least that of the blocks.

Clearly, $r_i \geqslant \lambda$, and there can be at most one variety for which $r_i = \lambda$, unless all blocks are identical, i.e. we have only one single block. If we can prove that AA' is not singular, i.e. that it is of rank v, then it follows that $b \geqslant v$, since b is the order of A', and AA' cannot have higher rank than either of its factors.

If AA' were singular, then the following system of equations would have a solution where not all x_i are zero:

$$r_1 x_1 + \lambda x_2 + ... + \lambda x_v = 0$$
$$\lambda x_1 + r_2 x_2 + ... + \lambda x_v = 0$$
$$- - - - - -$$
$$\lambda x_1 + \lambda x_2 + ... + r_v x_v = 0.$$

Then, by subtraction, $(r_j - \lambda) x_j = (r_i - \lambda) x_i$ for any i, j. If $r_i > \lambda$ for all i, then all x_i have the same sign and must, by virtue of the equations, vanish, because all coefficients are non-negative. On the other hand, if $r_i = \lambda$, then all $x_j = 0$ for $j \neq i$, and again from the equations we have $x_i = 0$ as well, because $r_i > 0$. Hence the system of equations has only the trivial solution $x_1 = x_2 = \dots = x_v = 0$, and AA' is not singular. (A weaker theorem which assumes that all variables are equally often repeated, was proved in [3] and in [70].)

For $\lambda = 1$ this has already been proved, by a different method, in [19], and the theorem referring to n lines through n points and their representation is a corollary of it. It is also shown in [19] that when $\lambda = 1$, then the number of blocks equals v if and only if the design is either a configuration, or such that one block contains $v-1$ varieties, and every other block contains two varieties, one of which is common with that block of $v-1$ varieties.

We proceed to prove a few theorems which give conditions for a design to be a configuration.

Theorem 4.3   If a design is of linked block type and balanced, then it is a symmetric configuration, provided $\lambda > 1$ ([52]).

Proof: Let r_i $(i = 1, \dots, v)$ be the number of replications of the ith variety, and k_j $(j = 1, \dots, b)$ the size of the jth block. Denote the incidence matrix by A, then

$$AA' = \begin{pmatrix} r_1 & \lambda & \dots & \lambda \\ \lambda & r_2 & \dots & \lambda \\ \multicolumn{4}{c}{----} \\ \lambda & \lambda & \dots & r_v \end{pmatrix} \quad \text{and} \quad A'A = \begin{pmatrix} k_1 & \mu & \dots & \mu \\ \mu & k_2 & \dots & \mu \\ \multicolumn{4}{c}{----} \\ \mu & \mu & \dots & k_b \end{pmatrix}.$$

Moreover, we see easily that

$$A \begin{pmatrix} k_1 \\ -- \\ k_b \end{pmatrix} = \begin{pmatrix} \lambda(v-1) + r_1 \\ --- \\ \lambda(v-1) + r_v \end{pmatrix} \quad \text{and} \quad A' \begin{pmatrix} r_1 \\ -- \\ r_v \end{pmatrix} = \begin{pmatrix} \mu(b-1) + k_1 \\ --- \\ \mu(b-1) + k_b \end{pmatrix}.$$

(For instance, the first variety appears altogether r_1 times, and where it appears, it does so altogether λ times with any of the other varieties; therefore the sum of the sizes of all those blocks where the first variety appears is $r_1 + \lambda(v-1)$.)

It follows that

$$A'A\begin{pmatrix} k_1 \\ - \\ k_b \end{pmatrix} = \begin{pmatrix} k_1 & \mu & \cdots & \mu \\ - & - & - & - \\ \mu & \mu & \cdots & k_b \end{pmatrix}\begin{pmatrix} k_1 \\ - \\ k_b \end{pmatrix}, \quad \text{and} \quad A'\begin{pmatrix} \lambda(v-1)+r_1 \\ - - - \\ \lambda(v-1)+r_v \end{pmatrix} =$$

$$\begin{pmatrix} \lambda(v-1)k_1 \\ - - - \\ \lambda(v-1)k_b \end{pmatrix} + \begin{pmatrix} \mu(b-1)+k_1 \\ - - - \\ \mu(b-1)+k_b \end{pmatrix}.$$

Comparison of the right-hand sides leads to b equations of the form

$$k_i^2 + \mu(k_1 + \cdots + k_b - k_i) = \lambda k_i(v-1) + \mu(b-1) + k_i,$$

i.e.

$$k_i^2 - k_i(\mu + \lambda v - \lambda + 1) + \mu(k_1 + \cdots + k_b - b + 1) = 0$$

for $i = 1, \ldots, b$.

We show that all k_i must be equal to the same root of this quadratic equation. Indeed, if k_1 and k_2, say, were equal to the two different roots of the equation, then, because the number of all varieties is at least $k_1 + k_2 - \mu$, we would have

$$v > k_1 + k_2 - \mu = (\mu + \lambda v - \lambda + 1) - \mu = \lambda(v-1) + 1,$$

and this is impossible if $\lambda > 1$.

Therefore all $k_i = k$, say and, because the design is balanced, all r_i equal $r = \lambda(v-1)/(k-1)$.

The determinants $|AA'|$ and $|A'A|$ are found to be $rk.(r-\lambda)^{v-1}$ and $(k+\mu b-\mu)(k-\mu)^{b-1}$ respectively. Both are positive, so that the rank of AA' is v, and that of $A'A$ is b. Therefore $b = v$, $r = k$, $\lambda = \mu$.

When $\lambda = 1$, then this result does not follow. We have then $v = k_1 + k_2 - 1$ (μ cannot exceed 1 in this case) and it can be shown that this is only possible if one block is of size $v-1$, and all others are of size 2.

On the other hand, if we assume that the design is a configuration, so that all blocks are of the same size by assumption, then it follows as above that the design is symmetric. We have thus proved the

Corollary: If a configuration is balanced and of linked block type, then it is symmetric.

This corollary follows also from the following lemma in [76]:

Lemma: If a symmetric matrix of order v has a $(v-1)$-fold latent root, and if the latent vector corresponding to the remaining root is $(1, \ldots, 1)$, then all diagonal elements of the matrix are equal, and all its off-diagonal elements are also equal.

84

Proof: Let the latent roots be x_1 (once) and x_2. Because B is symmetric, we can find an orthogonal matrix C such that

$$B = C' \begin{pmatrix} x_1 & 0 & 0 \\ 0 & x_2 & 0 \\ 0 & 0 & 0 \\ 0 & 0 & x_2 \end{pmatrix} C.$$

Because $(1,\dots,1)$ is a latent root, C can be chosen as $\begin{pmatrix} U \\ P \end{pmatrix}$, where $U = (1/\sqrt{v},\dots,1/\sqrt{v})$, and P is a $(v-1)$ by v matrix. Then $B = x_1 U'U + x_2 P'P$ has all diagonal elements equal to $[x_1 + (v-1)x_2]/v$, and all off-diagonal elements equal to $(x_1 - x_2)/v$. (These values are, of course, not necessarily integers.)

The corollary can now be derived as follows:

Let A be the incidence matrix of a configuration (b_k, v_r) (not necessarily balanced). Let AA' have a latent root rk with latent vector $(1,\dots,1)$, and let the only other non-zero latent root, $k-\mu$, say, have multiplicity $b-1$. Then the design is of linked block type, as will now be shown, as a preliminary to proving the lemma.

The non-zero latent roots of a product of matrices AB are the same as those of the product BA, with the same multiplicities; hence $k-\mu$ is also a latent root of multiplicity $b-1$ of $A'A$. Because the total of each column of AA' is rk, the latent vector belonging to this latent root is $(1,\dots,1)$. Now if x is a latent vector of AB, then xA is a latent vector of BA. Thus $(1,\dots,1)A = (k,\dots,k)$ is a latent vector of $A'A$, and so is $(1,\dots,1)$. It follows from the lemma proved above that $A'A$ has all its diagonal elements equal to k, and all its off-diagonal elements equal to μ. The design is of linked block type.

If the configuration is balanced, then the multiplicity $b-1$ equals $v-1$, i.e. if a configuration is balanced and of linked block type, then we have $b = v$: the configuration is symmetric.

Theorem 4.4 ([78]). If in a balanced design of index λ we have $v = b$ and the incidence matrix is not singular, and if moreover all varieties appear the same number (r) of times, then all blocks must have the same size, i.e. the design is a symmetric configuration.

Proof: The column sums of A' are all r, and the column sums of AA' are all $r + \lambda(v-1) = c$, say (a constant). Therefore $(1,\dots,1)AA' = (c,\dots,c)$.

Let the column sums of A be $s_1,\dots,s_b$, so that $(1,\dots,1)A =$

$(s_1, ..., s_b)$. Then $(1, ..., 1)AA' = (s_1, ..., s_b)A' = (c, ..., c)$. This system of equations in s_i is satisfied by $s_1 = ... = s_b = c/r$, and because A is not singular, this is the only solution. Hence the column sums of A are all equal to a constant which must be r. (Cf. [53], where a slightly more general theorem is proved.)

Theorem 4.5   (a) If v varieties are arranged in v blocks so that each variety appears in k blocks and any two varieties appear together in the same block $\lambda = k(k-1)/(v-1)$ times, then every block contains k varieties, and any two blocks have λ varieties in common.

(b) If v varieties are arranged in v blocks of k varieties each, and if any two blocks have $\mu = k(k-1)/(v-1)$ varieties in common, then each variety appears altogether k times and any two varieties appear μ times together in the same block ([18]).

(a) and (b) are dual to one another, and it is sufficient to prove (a).

Proof: It follows from the assumptions that

$$AA' = \begin{pmatrix} k & \lambda & \lambda \\ \lambda & k & \lambda \\ - & - & - & - & - \\ \lambda & \lambda & k \end{pmatrix}.$$

Introduce

$$B = \begin{array}{c|cc} -k & \sqrt{-\lambda} & \sqrt{-\lambda} \\ \hline \sqrt{-\lambda} & & \\ & A & \\ \sqrt{-\lambda} & & \end{array}$$

To evaluate

$$BB' = \left(\begin{array}{c|c} k^2 - \lambda v & -k\sqrt{-\lambda} + A\sqrt{-\lambda} \\ \hline (-k\sqrt{-\lambda} + A\sqrt{-\lambda})' & -\lambda J_v + AA' \end{array} \right)$$

we compute: $k^2 - \lambda v = k - \lambda$, $-k\sqrt{-\lambda} + A\sqrt{-\lambda} = (0, ..., 0)$, $-\lambda J_v + AA' = (k-\lambda)I_v$, so that $BB' = (k-\lambda)I_{v+1} = B'B$, and hence $AA' = A'A$. This proves the theorem.

Our next theorem deals with configurations.

Theorem 4.6   (a) If a configuration is balanced and symmetric, then it is of linked block type ([6]).

(b) If a configuration is of linked block type and symmetric, then it is balanced.

(a) and (b) are dual to one another, and it is sufficient to prove (a).

Proof: Denote the symmetric matrix of order v whose elements are all 1 by J_v. If $v = b$, $r = k$, $\lambda_{st} = \lambda$ for all s,t, then $AA' = (k-\lambda)I_v + \lambda J_v$, and
$$AJ_v = J_v A = kJ_v = rJ_v.$$

Hence, $A'A = A^{-1}(AA')A = A^{-1}(k-\lambda)I_v A + \lambda A^{-1}J_v A = (k-\lambda)I_v + \lambda J_v = AA'$. This proves the theorem, and also that $\lambda = \mu = k(k-1)/(v-1)$.

Taken in conjunction with the Corollary to Theorem 4.3, we see that if any two of the following properties hold in a configuration, namely linkage, balance, and symmetry, then the third holds as well.

We can now also show that a symmetric b.i.b.d. with $\lambda = 1$ is equivalent to a finite projective plane. We have $v = b = r^2 - r + 1$, and denoting $r-1$ by s, we have $v = b = s^2 + s + 1$, $r = k = s + 1$. By the above Theorem 4.6(a), we have also $\mu = 1$. Thus the blocks and varieties of a symmetric b.i.b.d. satisfy the conditions for lines and points of a finite projective plane.

The following theorem, which is weaker than 4.6, is proved in [43].

Theorem 4.7   Let there be v varieties arranged in b blocks so that each variety appears in the same number of blocks, any two blocks have at least μ varieties in common, and no block contains more than k varieties. Then $\mu(v-1) \leqslant k(k-1)$.

Proof: If $k = \mu$, then there can be only one block containing all varieties, and the relation to be proved is satisfied as an equation. If $k > \mu$, then select one particular variety, say x, and delete it from all the blocks in which it appears. Call these blocks of type A, and the others of type B. Let there be a blocks of the former, and b blocks of the latter type. We see that (i) each pair of blocks of different types has at least μ varieties in common, (ii) no block of type A has more than $k-1$ varieties, and (iii) no block of type B has more than k varieties.

Denote the elements other than x by $x_1, \ldots, x_{v-1}$ and let the number of blocks of type A containing x_i be r_i, and that of blocks of type B containing x_i be $r - r_i$. Then we have

from (i), $\displaystyle\sum_{i=1}^{v-1} r_i(r - r_i) \geqslant ab\mu$,

from (ii), $\displaystyle\sum_{i=1}^{v-1} r_i \leqslant a(k-1)$,   and from (iii), $\displaystyle\sum_{i=1}^{v-1}(r - r_i) \leqslant bk$.

Now $k(k-1)/\mu = a(k-1).bk/ab\mu$, and this is not smaller than

$$\sum_{i=1}^{v-1}(r-r_i)\sum_{i=1}^{v-1}r_i \Big/ \sum_{i=1}^{v-1}r_i(r-r_i) \;=\; \frac{\sum_{i=1}^{v-1}r_i\left[r(v-1)-\sum_{i=1}^{v-1}r_i\right]}{r\sum_{i=1}^{v-1}r_i-\sum_{i=1}^{v-1}r_i^2}.$$

It remains to be shown that this is not smaller than $v-1$, i.e. that

$$r(v-1)\,\Sigma r_i - (\Sigma r_i)^2 \;\geqslant\; r(v-1)\,\Sigma r_i - (v-1)\,\Sigma r_i^2,$$

or that $\qquad (v-1)\,\Sigma r_i^2 - (\Sigma r_i)^2 \;\geqslant\; 0.$

But this must be so, because the left-hand side equals $(v-1)$ times

$$\sum_{i=1}^{v-1}\left[r_i - \sum_{i=1}^{v-1}r_i/(v-1)\right]^2.$$

The equation $\mu(v-1) = k(k-1)$ can only hold if all r_i are equal. Then the inequality sign is everywhere replaced by the $=$ sign, every block of type A has precisely $k-1$ varieties, and every block of type B precisely k. Every block has precisely μ varieties in common with every other. This means, by Theorem 4.6, that the original design is a symmetric balanced configuration.

Theorems of a similar type are contained in [51].

The following theorems deal with the question of necessary properties of a matrix to be the incidence matrix of a configuration.

Theorem 4.8　If A is a v by v matrix with integer elements and such that $AA' = A'A = (k-\lambda)I_v + \lambda J_v = B$, say, where $k^2 - k = \lambda(v-1)$ and $0 < \lambda < k < v$, then either A or $-A$ is the incidence matrix of a symmetric balanced configuration.

Proof: First, note that $B^{-1} = (k^2 I_v - \lambda J_v)/(k^3 - \lambda k^2)$; this is easily verified by computing BB^{-1}, which turns out to be I_v.

From $A' = A^{-1}B$ it follows that $A'B^{-1}A = I_v$, and substituting for B^{-1} we obtain

$$A'A = (k-\lambda)I_v + \frac{\lambda}{k^2}A'J_v A. \qquad \cdot \qquad \cdot \qquad \cdot \quad (*)$$

From $A'A = B$ it follows that $\sum_i a_{ij}^2 = k$ for all i. Denoting $\sum_i a_{ij}$ by s_j, we have $A'J_v A = (s_i s_j)$ and hence, comparing diagonal entries on the two sides of $(*)$, $k = (k-\lambda) + \frac{\lambda}{k^2}s_j^2$, i.e. $s_j = k^2$ for all j. This means that s_j equals either k or $-k$.

If for any j we have $\sum_i a_{ij}^2 = k$ and $\sum_i a_{ij} = k$, then all a_{ij} (they are integers!) must be 0 or 1 for that j. If for any j we have $\sum_i a_{ij}^2 = k$, but $\sum_i a_{ij} = -k$, then all a_{ij} must be 0 or -1 for that j. But we cannot have both types within the same matrix A for different values of j, because all off-diagonal elements of AA' are $\lambda > 0$. This proves the theorem.

This proof is taken from [81]. The theorem is also proved in [79].

Theorem 4.9 Let A be a v by v matrix with integer elements whose column sums are non-negative, and let $AA' = (k-\lambda)I_v + \lambda J_v = B$, say, and $k^2 - k = \lambda(v-1)$. If k and λ are relatively prime, and if $k-\lambda$ is odd, then A is the incidence matrix of a symmetric balanced configuration.

We do not assume now, as we did in Theorem 4.8, that $AA' = A'A$, and therefore we cannot conclude that $\sum_i a_{ij}^2 = k$. However, we can again obtain $\sum_i a_{ij}^2 = (k-\lambda) + \frac{\lambda}{k^2} s_i^2$.

It follows that λs_i^2 must be divisible by k^2, for all j; since k and λ are relatively prime, s_j must be divisible by k, say $s_j = ku_j$, where u_j is an integer.

The s_j were assumed to be non-negative, but as a matter of fact none of the u_j, or of the s_j, can be zero, because if it were, then for such a j we should have

$$\sum_i a_{ij}^2 = k - \lambda, \qquad s_j^2 = \left(\sum_i a_{ij}\right)^2 = \sum_i a_{ij}^2 \quad (\text{mod } 2)$$

and therefore $k - \lambda = 0 \pmod 2$. But this contradicts the assumption that $k - \lambda$ was odd.

Comparing the elements on the two sides of $J_v AA' J_v = J_v B J_v$, we obtain $\sum_{j=1}^v s_j^2 = k^2 v$, and hence $\sum_{j=1}^v u_j^2 = v$. It follows, because all u_j are positive integers, that $u_j = 1$, and $s_j = k$, $\sum_i a_{ij} = k$ for all j. But this means, using the argument at the end of the proof of Theorem 4.8, that A is the incidence matrix of a symmetric balanced incomplete block design.

This proof is essentially taken from [79]. In [81] a slightly more general theorem is proved.

If $k - \lambda$ is even, then we cannot argue as we did above. In fact, in [81] a matrix is exhibited, with $k = 3$, $\lambda = 1$, which satisfies the other conditions of the theorem, but is not the incidence matrix of a symmetric balanced configuration. There also exist, of course, such incidence matrices with the same parameters, namely that for the points and lines of a $PG(2,2)$ (Fig. 1).

Theorem 4.10   If A is a square matrix of order v, with vk elements equal to 1, and the remaining $v^2 - vk$ elements are equal to 0, then $|A| \leqslant k(k-\lambda)^{\frac{1}{2}(v-1)}$, where $\lambda = k(k-1)/(v-1)$ and $0 < \lambda < k < v$. If equality holds, then A is the incidence matrix of a symmetric balanced configuration.

For the proof, we refer to [80].

If the number of elements 1 is not restricted as above, then A need not be such an incidence matrix, even if equality holds. The following is a simple counter-example, from [80].

567   1467   1257   1236   2347   1345   2456.

The determinant of the incidence matrix is $3(3-1)^{\frac{1}{2}(7-1)} = 24$, and yet the design is not a configuration.

Of the theorems relating various structural features we mention here just one more, proved in [65]:

If in a balanced configuration with parameters $v = 2\lambda + 2$, $b = 4\lambda + 2$, $k = \lambda + 1$, $n = 2\lambda + 1$, λ is even, then any two blocks have at least one variety in common. This is not necessarily true for odd λ. Moreover, no two blocks are the same. The latter fact is also true for odd λ, as proved in [86].

Resolvability

A configuration is called *resolvable* if its blocks can be grouped into sets of equal sizes, so that the blocks of each set contain between them all varieties once. If, moreover, any two blocks from different sets have the same number of varieties in common, then the configuration is called *affine resolvable*.

For instance, the $C(1,3,4,8)$ mentioned on page 75 is affine resolvable, and so is the $C(1,2,3,9)$ on page 73.

Let there be $r = b/n$ sets and n blocks in each of them. Choose a block B_0 and denote the number of varieties which this block has in common with the jth block of the ith set by n_{ij} ($j = 1, ..., n$; i referring to blocks in sets different from that of B_0). There are $n(r-1)$ different combinations of i and j.

Denote the mean of all n_{ij} by m. A variety which occurs in B_0 occurs in $r-1$ blocks of the other sets. This is true of all the k varieties in B_0, so that

$$m = \sum_i \sum_j n_{ij}/n(r-1) = k(r-1)/n(r-1) = k/n = kr/b = k^2/v$$

or $v = n^2 m$, $b = nr$, $k = nm$.

If the resolvable configuration is balanced, then we can derive relations between the parameters which are more stringent than those valid in the general case (cf. [8]). The $\binom{k}{2}$ pairs of varieties in B_0 appear in $(\lambda - 1)$ blocks of the other sets, so that $\sum_i \sum_j \binom{n_{ij}}{2} = (\lambda - 1)\binom{k}{2}$.

It follows that $\sum_i \sum_j n_{ij}^2 = k[(\lambda - 1)(k-1) + (r-1)]$.

We have
$$\sum_i \sum_j (n_{ij} - m)^2 = \sum_i \sum_j n_{ij}^2 - n(r-1)m^2$$
$$= k[\lambda(k-1) - k + r] - k^2(r-1)/n$$

and using the identities $\lambda = r(k-1)/(v-1)$, $b = rn$, $v = kn$, $vr = bk$ we obtain, after some straightforward arithmetic,
$$\sum_i \sum_j (n_{ij} - m)^2 = k(v-k)(b-v-r+1)/n(v-1).$$

The left-hand side cannot be negative, and therefore we have

Theorem 4.11   In a balanced resolvable configuration $b \geqslant v + r - 1$. This is a more stringent inequality than $b \geqslant v$ (Theorem 4.2). It is equivalent to $r - \lambda \geqslant k$, which can be seen as follows:

For $b \geqslant v + r - 1$ we can write $k(b - r) \geqslant k(v - 1)$, i.e.

$$rv - \lambda(v-1) - r \geqslant k(v-1), \text{ hence } (v-1)(r-\lambda) \geqslant k(v-1),$$

and, since $v > 1$, $r - \lambda \geqslant k$. (Cf. [96], and also [46].) In [70] the inequality $b \geqslant v + r - 1$ is proved for balanced resolvable designs with equal numbers of replications of all varieties, but not necessarily equal block sizes. If $\lambda = 1$, then $r(k-1) = v - 1 = nk - 1$, and because r is an integer, $n - 1$ must be divisible by $k - 1$, say $n = (k-1)t+1$. Hence

$$v = k[(k-1)t+1)], \quad r = kt+1, \quad b = (kt+1)[(k-1)t+1].$$

In [77] it is proved that $b \geqslant v + r - 1$ is a consequence of v being divisible by, but not equal to k. Let $v = nk$, $b = nr$. Then, from $\lambda(v-1) = r(k-1)$, we have $r = \lambda(nk-1)/(k-1) = \lambda n + \lambda(n-1)/(k-1)$. Therefore $\lambda(n-1)/(k-1) = g$, say, must be an integer.

Now consider $v + r - 1 = (n+g-1)(\lambda n + g)/g$. This is not larger than $b = n(\lambda n + g)$, and hence the result follows. (See also a proof in [55].)

This proof does not give any indication about when $b = v + r - 1$ holds. We return therefore to our first proof, to conclude: if the configuration is balanced and affine resolvable, then all n_{ij} are equal, therefore then, and only then $b = v + r - 1$. In this case, $rn = v + r - 1$, i.e. $r = (v-1)/(n-1)$, hence $b = n(v-1)/(n-1)$, $kn = v$, and $\lambda = (v-n)/n(n-1) = (r-1)/n = (k-1)/(n-1)$, i.e. $k = \lambda(n-1)+1$.

All n_{ij} equal $m = k^2/v = k/n = [\lambda(n-1)+1]/n$. This can only be an integer if λ is of the form $nt+1$, with t an integer ([77] and [55]). Then we obtain, by simple arithmetic,

$$\lambda = r - nm, \quad m = nt-t+1, \quad r = n^2t+n+1, \quad \text{and hence}$$
$$v = n^2(nt-t+1), \quad b = n(n^2t+n+1), \quad k = n(nt-t+1).$$

Example

A $EG(m,s)$ originates from a $PG(m,s)$ by omitting a particular $(m-1)$-flat. Take all the $(t-1)$-flats on the latter. Through each of these there pass s^{m-t} different t-flats, apart from those in the omitted $(m-1)$-flat.

Within the $EG(m,s)$ they form a pencil of s^{m-t} "parallel" t-flats of s^t points each. There are $V(m-1, t-1; s)$ such pencils, and the t-flats in any one of them contain between them all the s^m points of the $EG(m,s)$.

Considering the t-flats as blocks and the points as varieties, we obtain a resolvable design of $V(m-1, t-1; s)$ sets, with $v = s^m$, $b = V(m-1, t-1; s)s^{m-t}$, $r = V(m-1, t-1; s)$, $k = s^t$. Of course, $b \geqslant v+r-1$.

In particular, if $t = m-1$, then $b = v+r-1$, and any two blocks from different sets have s^{m-2} varieties in common.

When $m = 2$, then we have a $EG(2, s)$ and taking its lines as blocks and its points as varieties, we obtain an affine resolvable b.i.b.d. with $v = s^2$, $b = s^2+s$, $r = s+1$, $k = s$, $\lambda = 1$. This is also the case if we start from a finite projective plane which is not necessarily a $PG(2, s)$.

We prove now, following [32], that every design with these parameters is (affine) resolvable.

Choose one of the blocks, B_1 say, with varieties $a_1, ..., a_k$. Let c_1 be a variety not contained in B_1. Then there will be k blocks containing the pairs $c_1a_1, c_2a_2, ..., c_ka_k$ and, because of $\lambda = 1$, these blocks will be all different (since a_ia_j are already together in B_1).

There remains another block with c_1, which does not contain any of the a_i. Let this be block B_2, with varieties $c_1, ..., c_k$. For each of these c_i, B_2 is the only block which contains it but does not contain any a_i (because $r = k+1$, $\lambda = 1$).

Choosing again a variety not yet considered, and continuing in the same manner, we obtain blocks $B_2, ..., B_k$, none of them containing any of the a_i, and no two of them having any variety in common, because each is the only one for every one of its varieties which does not contain one of the a_i as well.

It follows that $B_1, B_2, ..., B_k$ contain k^2 varieties, all of them just once.

Each block determines uniquely such a set; there will be altogether $k+1$ sets, and the given design determines the sets uniquely; hence all b.i.b.d's with these parameters are isomorphic with one another.

In [88] it is shown how in a certain case a new symmetric b.i.b.d. can be constructed from the combination of a given symmetric b.i.b.d. and a given affine resolvable b.i.b.d.

Consider a b.i.b.d. with parameters $v_1 = b_1 = n^2t+n+1$, $r_1 = k_1 = nt+1$, $\lambda = t$, and write the blocks

$$B_1 \quad B_2 \qquad B_{n^2t+n+1}$$

Let there also be given an affine resolvable b.i.b.d. with n blocks in each set, and n^2t+n+1 sets (where n and t have the same values as in the symmetric b.i.b.d. previously mentioned) and write the blocks as

$$B_{1,1} \quad B_{1,2} \qquad B_{1,n^2t+n+1}$$
$$- \quad - \quad -$$
$$B_{n,1} \quad B_{n,2} \qquad B_{n,n^2t+n+1}$$

Its parameters are

$$v_2 = n^2(nt - t + 1), \quad b_2 = n(n^2t + n + 1)$$
$$r_2 = n^2t + n + 1, \quad k_2 = n(nt - t + 1), \quad \lambda = nt + 1.$$

Let the varieties of the two designs be denoted by different symbols, so that we have altogether $v = v_1 + v_2 = n(n^2t+n+1)$ varieties.

The new design is constructed as follows:

Form $n(n^2t+n+1)$ blocks by combining, for all i, the block B_i and the block $B_{1,i}$, the blocks B_i and $B_{2,i}, ...,$ the blocks B_i and $B_{n,i}$. Moreover, form a further block containing all the varieties of the first design. Each of the new $b = n(n^2t+n+1)+1$ blocks will contain $k = v_1 = k_1 + k_2 = n^2t+n+1$ varieties. It is easily seen that each variety is repeated $r = nr_1 + 1 = n^2t+n+1$ times, and that the resulting configuration is balanced, with index $\lambda = \lambda_2 = nt+1$.

We now prove the following

Theorem 4.12   If a resolvable configuration has any two of the following properties, namely (i) it is affine, (ii) it is balanced, (iii) $b = v+r-1$, then it has the third as well.

Proof: We concluded from Theorem 4.11 that (iii) follows from (i) and (ii), and that (i) follows from (ii) and (iii). It remains therefore merely to prove that (ii) follows from (i) and (iii): if a configuration is affine resolvable and $b = v+r-1$, then it is balanced.

For the purpose of the proof we construct a matrix as follows: let an affine resolvable configuration with r sets of $n = b/r$ blocks in each be given. Let there correspond to each of the sets a row, and to each variety a column. Within each set, order the blocks and mark them $0, 1, ..., n-1$. The entry in the ith row and jth column of the matrix is the number of the block in which the jth variety appears in the ith set.

Example

From the $C(1,2,3,9)$ quoted above we obtain the matrix

$$\begin{matrix} 0 & 0 & 0 & 1 & 2 & 1 & 2 & 1 & 2 \\ 0 & 1 & 2 & 0 & 0 & 1 & 2 & 2 & 1 \\ 0 & 1 & 2 & 2 & 1 & 0 & 0 & 1 & 2 \\ 0 & 1 & 2 & 1 & 2 & 2 & 1 & 0 & 0 \end{matrix}$$

Take any two rows of the matrix, corresponding to two selected sets, say i_1 and i_2. Any pair of numbers, say u and v, will appear in row i_1 and i_2 respectively, in the same column as often as the number of common varieties in block u of set i_1 and block v of set i_2. In an affine resolvable configuration this number will be the same for any pair u,v, and for any pair i_1,i_2. (A matrix with this property is called an "orthogonal array of strength 2", and such and more general arrays are studied in [103].)

If the same number appears in the same row of two selected columns, then this means that the two varieties, to which the columns correspond, appear together in the same block. In a balanced configuration this will happen equally often, whatever two columns we take.

To prove Theorem 4.12, we have to show that in an orthogonal array of strength 2 the number of pairs of equal symbols in the same row, in any selected pair of columns, is independent of which pairs of columns we have selected. We prove this, following the proof given in [67].

Choose a n by n matrix Q whose first row consists of 1's only and which has the orthogonal property $QQ' = nI_n$. Number the columns of Q from 0 to $n-1$. In the array of r rows and v columns derived from the configuration, replace each symbol by the corresponding column of Q, but omitting the top 1. Finally, add to the transformed array a single top row of v 1's.

Example

Writing a for $\sqrt{3/2}$, b for $\sqrt{1/2}$ and c for $\sqrt{2}$, we take for Q the

matrix $\begin{pmatrix} 1 & 1 & 1 \\ a & 0 & -a \\ b & -c & b \end{pmatrix}$ From the array of the $C(1,2,3,9)$

we obtain $A = $
$$\begin{pmatrix}
1 & 1 & 1 & 1 & 1 & 1 & 1 & 1 & 1 \\
a & a & a & 0 & -a & 0 & -a & 0 & -a \\
b & b & b & -c & b & -c & b & -c & b \\
a & 0 & -a & a & a & 0 & -a & -a & 0 \\
b & -c & b & b & b & -c & b & b & -c \\
a & 0 & -a & -a & 0 & a & a & 0 & -a \\
b & -c & b & b & -c & b & b & -c & b \\
a & 0 & -a & 0 & -a & -a & 0 & a & a \\
b & -c & b & -c & b & b & -c & b & b
\end{pmatrix}$$

We obtain a matrix A of order $r(n-1)+1$ by v, and if $b = rn = v + r-1$, then this matrix is square. We assume this to be the case. Then $AA' = vI_v$, and hence also $A'A = vI_v$. This follows from the orthogonal property of Q, remembering that in the original array all numbers appeared the same number of times in any row, and that any number in any row appeared in the same column with all other numbers in any other row.

By the orthogonal property of Q, a pair of unequal symbols in the same row in any two columns of the original array contributes -1 to the scalar product of the two corresponding columns of A (since the top 1 was omitted at the replacement), and a pair of equal symbols contributes $n-1$. Hence if there are in the two columns of the array λ pairs of equal symbols in the same row, then we have (remembering the added top row 1's in A)

$$1 + (-1)(r-\lambda) + (n-1)\lambda = 0;$$

hence

$$\lambda = (r-1)/n = (v-n)/n(n-1).$$

This value is the same for any pair of columns, which proves the last part of Theorem 4.12.

A resolvable balanced configuration with $\lambda = 1$ solves a generalization of Kirkman's schoolgirl problem (see [47]):

15 girls go for a walk every day of the week, three abreast. How can these walks be arranged so that during the seven days of a week any pair of girls walks once in the same line?

This problem is seen to be that of $v = 15$, $b = 35$, $k = 3$, $r = 7$, $\lambda = 1$. (Cf. Chapter X of [75] where it is stated that solutions exist for all $v = 6t+3$, $k = 3$, $r = 3t+1$.) We give here a possible arrangement:

1	2	3		2	4	6		4	3	7		3	6	5
4	12	8		3	9	13		6	14	10		7	11	12
7	9	14		5	14	11		1	11	8		2	8	13
6	11	13		7	8	10		5	13	12		1	10	9
5	10	15		1	12	15		2	9	15		4	14	15

	5	1	4		6	7	1		7	5	2
	2	10	11		5	8	9		1	13	14
	6	12	9		4	13	10		3	10	12
	3	14	8		2	12	14		4	9	11
	7	13	15		3	11	15		6	8	15

Kirkman was actually interested in the existence of resolvable configurations such that no q-plet appeared more than once within the same block, and investigated the question for which v and k such configurations could exist. He showed that this was the case whenever $q = 2$, $k = 3$, $v = 5.3^{m+1}$, and hence $b = \frac{1}{2}5.3^m(5.3^{m+1}-1)$, $r = \frac{1}{2}(5.3^{m+1}-1)$. The solutions are, of course, Steiner triple systems. 5.3^{m+1} is always of the form $6t+3$.

Intersection and section

Two methods of deriving designs from configurations, called respectively "block intersection" and "block section", will now be described.

Block intersection

From a given configuration C omit one block, and retain in the other blocks only those varieties which were contained in the omitted block. There remain $v' = k$ varieties, and each of these is repeated $r' = r-1$ times.

If the original configuration was of linked block type with index μ, then the new design has $b' = b-1$ blocks, each of size $k' = \mu$; if the original configuration was pairwise balanced with index λ, say, then the new one is also balanced, with index $\lambda-1$.

If the original design was affine resolvable, then $b' = b(r-1)/r$ and $k' = k^2/v$.

Block section (residual design)

Omit one block and all its varieties from all blocks. There remain $v' = v-k$ varieties and $b' = b-1$ blocks. Each variety is repeated $r' = r$ times.

If the original configuration was of linked block type, then each block of the new design is of size $k' = k-\mu$; if it was balanced, the new one is also balanced, with the same index as before.

It will be noticed that the residual design of a $PG(m,s)$ is the $EG(m,s)$.

The procedure which is opposite to that of block section is called "embedding".

Embedding

We are interested in finding conditions for a b.i.b.d. to be adjoinable, i.e. that it can be considered to have been derived by block section from a symmetric b.i.b.d., in which it is then said to be embedded.

A symmetric b.i.b.d. is balanced as well as of linked block type. Let its parameters be v', b', r', k', λ'. In a design obtained from it by block section we have

$$v = v' - k', \quad b = b' - 1, \quad r = k', \quad k = k' - \lambda', \quad \lambda = \lambda'.$$

It is therefore a b.i.b.d.

From $\lambda'(v'-1) = r'(k'-1)$ we have $\lambda b = k(k+\lambda-1)$, and from $\lambda'(v'-1) = k'(k'-1)$ we have

$$\lambda(v+k+\lambda-1) = (k+\lambda)(k+\lambda-1), \quad \text{i.e.} \quad \lambda v = k(k+\lambda-1).$$

These relationships must hold between the parameters of an adjoinable b.i.b.d. We derive further necessary properties.

Let those varieties of the original symmetric b.i.b.d. which were omitted be $x_1, \ldots, x_{k'}$, and let all those blocks of the new b.i.b.d.

which contained one of them, say x_j, before its deletion, form a set S_j. The $k' = k+\lambda$ sets for the different j overlap, unless $\lambda = 1$. They have the following properties.

(i)   Each S_j consists of $k+\lambda-1$ blocks.

In the original b.i.b.d. each variety was repeated $r' = k+\lambda$ times, and one block containing it was omitted.

(ii)   The blocks of S_j contain all (retained) varieties λ times between them.

Originally x_j appeared λ times in the same block with every other variety, and hence with every retained variety. This must have occurred in the retained blocks.

(iii)   Any two sets S_j and S_k have $\lambda-1$ blocks in common.

x_j and x_k appeared λ times together in the original blocks, and one of them has been omitted.

(iv)   Each block of the new b.i.b.d. appears in λ sets S_j.

The omitted block had λ varieties in common with every other original block, so that each of the latter lost λ varieties and appears now in all corresponding S_j.

A b.i.b.d. can only be adjoinable if its blocks can be grouped into sets with properties (i)-(iv), and if the parameters are related as shown. On the other hand, these relations, and the possibility of forming the blocks into such sets, are sufficient for the b.i.b.d. to be adjoinable. We see this by adjoining to it a new block consisting of $k+\lambda$ new varieties $x_1, \ldots, x_{k+\lambda}$ and adding the variety $x_u, x_v, \ldots$ to all blocks which appear in the sets $S_u, S_v, \ldots$. It will be seen that we obtain a symmetric b.i.b.d.

If, in particular, $\lambda = 1$, then the properties (i)-(iv) mean that the b.i.b.d. is resolvable, and since the relations between the parameters are then $v = k^2$, $b = k^2+k$, $r = k+1$ so that $b = v+r-1$, the configuration must be affine resolvable. But we have seen that a b.i.b.d. with those relationships between its parameters is necessarily affine resolvable, and hence it is *adjoinable*, i.e. it can be embedded into a symmetric b.i.b.d. with $\lambda = 1$. This embedding is (apart from isomorphisms) obviously unique.

When $\lambda = 2$, then again every b.i.b.d. with the appropriate relations between its parameters, i.e. in this case $2v = k^2+k$, $2b = (k+1)(k+2)$, $r = k+2$, is uniquely embeddable in a symmetric b.i.b.d.

This was proved in [32] and a constructive proof is given in [89].

For $\lambda = 3$ embedding might be impossible even when the parameters satisfy the necessary conditions, namely

$$3b = (k+2)(k+3), \quad 3v = k(k+2), \quad r = k+3.$$

Bhattacharya has given the following counter-example with $v = 16$, $b = 24$, $k = 6$, $r = 9$ in [5].

	1	2	7	8	14	15	1	4	7	8	11	16
	3	5	7	8	11	13	2	4	8	10	12	14
	2	3	8	9	13	16	5	6	8	10	12	13
	3	5	8	9	12	14	* 1	6	8	10	12	13
* 1	6	7	9	12	13		1	2	3	11	12	15
	2	5	7	10	13	15	2	6	7	9	14	16
	3	4	7	10	12	16	1	4	5	13	14	16
	3	4	6	13	14	15	2	5	6	11	12	16
	4	5	7	9	12	15	1	3	9	10	15	16
	2	4	9	10	11	13	4	6	8	9	11	15
	3	6	7	10	11	14	1	5	9	10	11	14
+	1	2	3	4	5	6	+ 11	12	13	14	15	16

The blocks marked * have four varieties in common, and those marked + have none. It is therefore impossible to embed them in a symmetric b.i.b.d. with $\lambda = 3$.

Other designs

In [60], Nair and Rao have defined intra- and inter-group b.i.b.d's as designs of v varieties in b blocks of k varieties each, and such that v_i of the varieties are repeated r_i times. Moreover, two varieties, one of them repeated r_i and the other r_j times, appear in the same block λ_{ij} times, whichever the two varieties are.

Numerous designs, too many to quote them all here, have been used for special purposes. An excellent summary, with an extensive bibliography, will be found in [59].

Latin squares and orthogonal arrays are treated in great detail in [103].

EXERCISES

(1) Construct an affine resolvable balanced incomplete block design with $v = 9$, $b = 12$, $r = 4$, $k = 3$ from a $EG(2,3)$.

(2) Show that the following balanced incomplete block design is adjoinable: $v = 6$, $b = 10$, $r = 5$, $k = 3$, $\lambda = 2$:

| 1 2 3 | 1 4 6 | 3 4 5 | 1 2 5 | 2 4 5 |
| 1 5 6 | 1 3 4 | 3 5 6 | 2 4 6 | 2 3 6 |

(3) Describe the configuration of lines and points of a $PG(3,2)$. Is it resolvable?

(4) Construct resolvable designs from a $EG(3,2)$.

(5) Construct an affine resolvable design from a $EG(3,3)$.

(6) The t-flats in a $PG(m,s)$ form a $C(\lambda, 2, k, v)$. Do they also form a $C(\lambda_\mu, \mu, k, v)$ for $2 < \mu < t$?

(7) Derive the b.i.b.d. for $v = 9$, i.e. $t = 1$, from Bose's method given on page 73.

SOLUTIONS TO EXERCISES

Chapter I

(1)   Call the matrices 1, 2, 3, 4, 5, 6 in the order given. The permutation group is

$$I, \quad (123)\,(465), \quad (132)\,(456), \quad (14)\,(25)\,(36)$$
$$(15)\,(26)\,(34), \quad (16)\,(24)\,(35).$$

Multiplication table:

	1	2	3	4	5	6
1	1	2	3	4	5	6
2	2	3	1	6	4	5
3	3	1	2	5	6	4
4	4	5	6	1	2	3
5	5	6	4	3	1	2
6	6	4	5	2	3	1

(2)   Call the matrices 1, 2, 3, 4, 5, 6 in the order given. The permutation group is cyclic:

$$I, \quad (153426), \quad (132)\,(465), \quad (14)\,(25)\,(36)$$
$$(123)\,(456), \quad (162435).$$

(3)   Use the function $x^2 + x + 1$. Call the elements $0, 1, y, z = 1+y$.

Multiplication table:

	0	1	y	z
0	0	0	0	0
1	0	1	y	z
y	0	y	z	1
z	0	z	1	y

Addition table:

	0	1	y	z
0	0	1	y	z
1	1	0	z	y
y	y	z	0	1
z	z	y	1	0

(4) $2^2 = 4$, $2^4 = 16$, $2^6 = 7$, $2^8 = 9$, $2^{10} = 17$, $2^{12} = 11$,
$2^{14} = 6$, $2^{16} = 5$, $2^{18} = 1$.

Difference set: 1, 4, 5, 6, 7, 9, 11, 16, 17.

Chapter II

(1)   Line AB $x_0 + yx_1 + x_2 = 0$ CD $x_0 \qquad\qquad + (1+y)x_2 = 0$
AC $x_0 + x_1 \qquad = 0$ BD $x_0 + (1+y)x_1 \qquad + yx_2 = 0$
$AD \qquad x_1 + yx_2 = 0$ BC $\qquad x_1 + (1+y)x_2 = 0.$

The diagonal points are $(1, y, y)$, $(1, 1, 1)$, and $(1, 0, 0)$, and they are all on the line $x_1 + x_2 = 0$.

(2)   The lines $Aa \ldots x_1 = 0$, $Bb \ldots x_2 = 0$, $Cc \ldots x_1 + x_2 = 0$ pass through the point $P(1, 0, 0)$.

The lines $AB \ldots x_0 + x_1 + x_2 = 0$ and $ab \ldots 2x_0 + x_1 + x_2 = 0$
intersect in the point $(0, 2, 3)$, say Q.

The lines $AC \ldots 2x_0 + 3x_1 + 2x_2 = 0$ and $ac \ldots 2x_0 + x_2 = 0$
intersect in the point $(1, 4, 3)$, say R.

The lines $BC \ldots x_0 + x_1 + 3x_2 = 0$ and $bc \ldots x_0 + 3x_1 + x_2 = 0$
intersect in the point $(1, 1, 1)$, say S.

The points Q, R, and S lie on the line $3x_0 + x_1 + x_2 = 0$.

(3)   The remaining lines through the point A are

$AE \ldots x_0 + yx_2 = 0$ and $AF \ldots x_0 + (1+y)x_1 + (1+y)x_2 = 0.$

The intersections are

with	BC	$(1, y, 1+y)$	$(1, 1+y, 1)$
	BD	$(1, 0, 1+y)$	$(1, y, 0)$
	CD	$(0, 1, 0)$	$(1, 0, y)$

Chapter II, (3) continued

Further points are

$$E(1,\ 1+y,\ 1+y) \qquad\qquad F(0,\ 1,\ 1)$$

(4)   We might take the points A, B, C, D, and E of Exercise (3). Of the five lines through A, the line $AF \ldots x_0 + (1+y)x_1 + (1+y)x_2 = 0$ is a tangent. The other tangents are:

$$
\begin{array}{llrl}
\text{Through} & B & x_0 & = 0 \\
& C & x_0 + yx_1 + yx_2 & = 0 \\
& D & x_0 + x_1 + x_2 & = 0 \\
& E & x_1 + x_2 & = 0.
\end{array}
$$

All these pass through the point $F(0,\ 1,\ 1)$, the nucleus.

(5)   Take the points $A(1,\ 0,\ 0)$, $B(1,\ 1,\ 0)$, $C(1,\ 2,\ 1)$, $D(1,\ 2,\ 2)$, $E(1,\ 1,\ 2)$. No three of these are on the same line. If we list all points of the plane and exclude those on any diagonal of $ABCDE$, then only one point is left, say $F(1,\ 0,\ 1)$. The lines through the six points of the conic are:

$$
\begin{array}{lllrl}
\text{Through} & A & \text{to} & B & x_2 = 0 \\
& & & C & x_1 + 3x_2 = 0 \\
& & & D & x_1 + 4x_2 = 0 \\
& & & E & x_1 + 2x_2 = 0 \\
& & & F & x_1 = 0 \\
& & \text{tangent} & & x_1 + x_2 = 0 \\[6pt]
\text{Through} & B & \text{to} & C & x_0 + 4x_1 + x_2 = 0 \\
& & & D & x_0 + 4x_1 + 3x_2 = 0 \\
& & & E & x_0 + 4x_1 = 0 \\
& & & F & x_0 + 4x_1 + 4x_2 = 0 \\
& & \text{tangent} & & x_0 + 4x_1 + 2x_2 = 0 \\[6pt]
\text{Through} & C & \text{to} & D & x_0 + 2x_1 = 0 \\
& & & E & x_0 + 3x_1 + 3x_2 = 0 \\
& & & F & x_0 + 4x_2 = 0 \\
& & \text{tangent} & & x_0 + x_1 + 2x_2 = 0
\end{array}
$$

$$\text{Through } D \text{ to } E \quad x_0 + \qquad 2x_2 = 0$$
$$F \quad x_0 + 3x_1 + 4x_2 = 0$$
$$\text{tangent} \quad x_0 + x_1 + x_2 = 0$$

$$\text{Through } E \text{ to } F \quad x_0 + x_1 + 4x_2 = 0$$
$$\text{tangent} \quad x_0 + 2x_1 + x_2 = 0$$

$$\text{Tangent through } F \quad x_0 + 2x_1 + 4x_2 = 0.$$

(6)
$$x_0^2 + x_1^2 + y(x_0 x_1 + x_0 x_2 + x_2^2) = 0.$$

The tangents intersect in the following points, which are all different:

	B	C	D	E	F
A	(1, 2, 3)	(1, 1, 4)	(0, 1, 4)	(1, 4, 1)	(1, 3, 2)
B		(1, 0, 2)	(1, 3, 1)	(0, 2, 1)	(1, 4, 4)
C			(1, 4, 0)	(1, 3, 3)	(0, 1, 2)
D				(1, 0, 4)	(1, 1, 3)
E					(1, 2, 0)

Chapter III

(1) A $PG(4,3)$ has $(3^5 - 1)/(3-1) = 121$ points

$$\frac{(3^5 - 1)(3^4 - 1)}{(3^2 - 1)(3-1)} = 1210 \text{ lines,}$$

$$\frac{(3^5 - 1)(3^4 - 1)(3^3 - 1)}{(3^3 - 1)(3^2 - 1)(3-1)} = 1210 \text{ planes, and}$$

$$\frac{(3^5 - 1)(3^4 - 1)(3^3 - 1)(3^2 - 1)}{(3^4 - 1)(3^3 - 1)(3^2 - 1)(3-1)} = 121 \text{ 3-flats.}$$

There are 4 points on a line; 13 points and 13 lines on a plane; 40 points, 130 lines and 40 planes in a 3-flat; 40 lines, 130 planes and 40 3-flats through a point; 13 planes and 13 s-flats through a line; 4 3-flats through a plane.

A line is defined by 2 points, or as an intersection of 3 3-flats; a plane is defined by 3 points, or as an intersection of 2 3-flats.

(2) $x_1 = 0, \quad x_3 = 0, \quad x_1 + x_3 = 0, \quad x_1 + 2x_3 = 0.$

Chapter III, (2) continued

Any two of these are independent, but no three of them. This is seen by inspecting the following matrix, which is of rank 2:

$$\begin{pmatrix} 0 & 1 & 0 & 0 & 0 \\ 0 & 0 & 0 & 1 & 0 \\ 0 & 1 & 0 & 1 & 0 \\ 0 & 1 & 0 & 2 & 0 \end{pmatrix}$$

(3) $$x_0 y_1 + x_0 y_2 + x_1 y_0 + x_1 y_2 + x_2 y_0 + x_2 y_1 = 0.$$

	The points $(y_0 y_1 y_2)$	are contained in the lines	together with
a	(1, 0, 0)	$x_1 + x_2 = 0$	d, g
b	(1, 0, 1)	$x_0 + x_2 = 0$	d, e
c	(1, 1, 0)	$x_1 + x_0 = 0$	d, f
d	(1, 1, 1)	no lines	
e	(0, 1, 0)	$x_2 + x_0 = 0$	b, d
f	(0, 0, 1)	$x_0 + x_1 = 0$	c, d
g	(0, 1, 1)	$x_2 + x_1 = 0$	a, d

(4)	Points	contained in lines	which contain also
a	(0, 1, 0)	$x_0 \quad + x_2 = 0$	i, k, m
b	(0, 0, 1)	$2x_0 + x_1 \quad = 0$	f, h, k
c	(0, 1, 1)	$x_0 + x_1 - x_2 = 0$	g, j, k
d	(0, 1, 2)	$2x_1 - x_2 = 0$	e, k, l
e	(1, 0, 0)	$x_1 - 2x_2 = 0$	d, k, l
f	(1, 1, 0)	$-x_0 + x_1 \quad = 0$	b, h, k
g	(1, 0, 1)	$x_0 + x_1 - x_2 = 0$	c, j, k
h	(1, 1, 1)	$x_0 + 2x_1 \quad = 0$	b, f, k
i	(1, 0, 2)	$-2x_0 \quad + x_2 = 0$	a, m, k
j	(1, 2, 0)	$2x_0 - x_1 + x_2 = 0$	c, g, k
k	(1, 1, 2)	no lines	
l	(1, 2, 1)	$-2x_1 + x_2 = 0$	d, e, k
m	(1, 2, 2)	$-2x_0 \quad + x_2 = 0$	a, i, k

– – –

Chapter III, continued

(5)
```
1  0  0  1        1  0  0  0
1  1  0  1        1  1  0  0
1  0  1  1        1  0  1  0
1  1  1  1        1  1  1  0
```

These are co-ordinates of the vertices of a cube.

There are no three of these on a line. On the other hand, there are sets of four that are on the same plane.

(6) $s = 5$ 1   2   3   4   5
                 7   8   9 10 11
                13 14 15 16 17   6, 12, and 18 do not appear.
                19 20 21 22 23

$s = 7$ 1   2   3   4   5   6   7
          9 10 11 12 13 14 15
         17 18 19 20 21 22 23
         25 26 27 28 29 30 31
         33 34 35 36 37 38 39
         41 42 43 44 45 46 47

8, 16, 24, 32, and 40 do not appear.

Chapter IV

(1)   Blocks:

x_1 = 0 with points   00 01 02,

$\qquad\qquad\qquad\qquad\qquad x_2 = 0$ with points   00 10 20

$1 + x_1$ = 0 with points   20 21 22,

$\qquad\qquad\qquad\qquad 1 + \quad x_2 = 0$ with points   02 12 22

$1 + 2x_1$ = 0 with points   10 11 12,

$\qquad\qquad\qquad\qquad 1 + \quad 2x_2 = 0$ with points   01 11 21

$2x_1 + x_2 = 0$ with points   00 11 22,

$\qquad\qquad\qquad\qquad x_1 + \quad x_2 = 0$ with points   00 12 21

$1 + \quad x_1 + x_2 = 0$ with points   11 20 02,

$\qquad\qquad\qquad\qquad 1 + \quad x_1 + 2x_2 = 0$ with points   20 01 12

$1 + 2x_1 + x_2 = 0$ with points   10 02 21,

$\qquad\qquad\qquad\qquad 1 + 2x_1 + 2x_2 = 0$ with points   10 01 22

Chapter IV, continued

(2)

	s_1	s_2	s_3	s_4	s_5
	123	123	245	356	146
	245	156	156	246	345
	356	246	134	134	236
	146	345	236	125	125

from this

```
1  2  3  7  8
2  4  5  7  9        1  5  6  8  9
3  5  6  7  10       2  4  6  8  10
1  4  6  7  11       3  4  5  8  11

1  3  4  9  10
2  3  6  9  11       1  2  5  10  11

      7  8  9  10  11
```

(3)   Points: $A(1, 0, 0, 0)$, $B(1, 1, 0, 0)$, $C(1, 0, 1, 0)$, $D(1, 0, 0, 1)$
             $E(1, 1, 1, 0)$, $F(1, 1, 0, 1)$, $G(1, 0, 1, 1)$, $H(1, 1, 1, 1)$
             $I(0, 1, 0, 0)$, $J(0, 0, 1, 0)$, $K(0, 0, 0, 1)$, $L(0, 1, 1, 0)$
             $M(0, 1, 0, 1)$, $N(0, 0, 1, 1)$, $P(0, 1, 1, 1)$.

The configuration (which is seen to be resolvable), is as follows:

```
J K N    A F M    B C L    G H I    D E P
I K M    A C J    E F N    D H L    B G P
I J L    A D K    B H N    E G M    C F P
A B I    J M P    E H K    F G L    C D N
K L P    C E I    F H J    B D M    A G N
I N P    B F K    A E L    C H M    D G J
C G K    D F I    B E J    A H P    I M N
```

(4) $t = 1$ $t = 2$
    7 points   7 lines
                 on the plane at infinity.
                    There are

    $7 - 3 = 4$ lines $3 - 1 = 2$ planes

                through each of these with

Chapter IV, (4) continued

2	4
finite points on each	
line	plane.

The pencils are as follows:

x_1 = 0, 1	x_2 = 0, 1	x_1 = 0, 1
x_1 = 0, 1	x_3 = 0, 1	x_2 = 0, 1
x_1 = 0, 1	x_2+x_3 = 0, 1	x_3 = 0, 1
x_2 = 0, 1	x_3 = 0, 1	$x_1+\ x_3$ = 0, 1
x_2 = 0, 1	$x_1+\ x_3$ = 0, 1	x_1+x_2 = 0, 1
x_3 = 0, 1	x_1+x_2 = 0, 1	x_2+x_3 = 0, 1
x_1+x_2 = 0, 1	x_2+x_3 = 0, 1	$x_1+x_2+x_3$ = 0, 1

$v = 8, b = 28, r = 7, k = 2$

$(x_1, x_2) = (0,0)\ (0,1)\ (1,0)\ (1,1)$
| 000 | 010 | 100 | 110 |
| 001 | 011 | 101 | 111 |

$(x_1, x_3) = (0,0)\ (0,1)\ (1,0)\ (1,1)$
| 000 | 001 | 100 | 101 |
| 010 | 011 | 110 | 111 |

$(x_1, x_2+x_3) = (0,0)\ (0,1)\ (1,0)\ (1,1)$
| 000 | 001 | 100 | 101 |
| 011 | 010 | 111 | 110 |

$(x_2, x_3) = (0,0)\ (0,1)\ (1,0)\ (1,1)$
| 000 | 001 | 010 | 011 |
| 100 | 101 | 110 | 111 |

$(x_2, x_1+x_3) = (0,0)\ (0,1)\ (1,0)\ (1,1)$
| 000 | 001 | 010 | 011 |
| 101 | 100 | 111 | 110 |

$(x_3, x_1+x_2) = (0,0)\ (0,1)\ (1,0)\ (1,1)$
| 000 | 010 | 001 | 011 |
| 110 | 100 | 111 | 101 |

$v = 8, b = 14, r = 7, k = 4$

$x_1 = 0$ 000 001 010 011
$\ \ \ = 1$ 100 101 110 111

$x_2 = 0$ 000 001 100 101
$\ \ \ = 1$ 010 011 110 111

$x_3 = 0$ 000 010 100 110
$\ \ \ = 1$ 001 011 101 111

$x_1+x_3 = 0$ 000 010 101 111
$\ \ \ \ \ \ \ = 1$ 001 011 100 111

$x_1+x_2 = 0$ 000 001 110 111
$\ \ \ \ \ \ \ = 1$ 010 011 100 101

$x_2+x_3 = 0$ 000 100 011 111
$\ \ \ \ \ \ \ = 1$ 001 101 010 110

Chapter IV, (4) continued

$(x_1 + x_2, x_2 + x_3) = (0, 0) \ (0, 1) \ (1, 0) \ (1, 1)$

000	011	001	010	$x_1 + x_2 + x_3 = 0$	000	011	101	110
111	100	110	101	$= 1$	100	010	001	111

       resolvable   affine resolvable.

(5)   There are 13 lines on the plane at infinity, and through each there are $4 - 1 = 3$ finite planes. On each of these there are 9 finite points. We obtain a design with $v = 27$, $b = 39$, $r = 13$, $k = 9$.

The pencils are as follows:

$x_1 =$			$x_2 =$			$x_3 =$		
0	1	2	0	1	2	0	1	2
000	100	200	000	010	020	000	001	002
001	101	201	001	011	021	010	011	012
002	102	202	002	012	022	020	021	022
010	110	210	100	110	120	100	101	102
011	111	211	101	111	121	110	111	112
012	112	212	102	112	122	120	121	122
020	120	220	200	210	220	200	201	202
021	121	221	201	211	221	210	211	212
022	122	222	202	212	222	220	221	222

$x_1 + x_2 =$			$x_1 + x_3 =$			$x_2 + x_3 =$		
0	1	2	0	1	2	0	1	2
000	010	020	000	001	002	000	001	002
001	011	021	010	011	012	100	101	102
002	012	022	020	021	022	200	201	202
120	100	200	102	100	200	012	010	020
121	101	201	112	110	210	112	110	120
122	102	202	122	120	220	212	210	220
210	220	110	201	202	101	021	022	011
211	221	111	211	212	111	121	122	111
212	222	112	221	222	121	221	222	211

Chapter IV, (5) continued

$x_1 + 2x_2 =$		
0	1	2
000	100	200
001	101	201
002	102	202
110	020	010
111	021	011
112	022	012
220	210	120
221	211	121
222	212	122

$x_1 + 2x_3 =$		
0	1	2
000	100	001
010	110	011
020	120	021
101	002	102
111	012	112
121	022	122
202	201	200
212	211	210
222	221	220

$x_2 + 2x_3 =$		
0	1	2
000	010	020
100	110	120
200	210	220
011	002	001
111	102	101
211	202	201
022	021	012
122	121	112
222	221	212

$x_1 + x_2 + x_3 =$		
0	1	2
000	001	002
012	010	011
021	022	020
111	112	110
102	100	101
120	121	122
210	211	212
201	202	200
222	220	221

$x_1 + x_2 + 2x_3 =$		
0	1	2
000	002	001
011	010	012
022	021	020
112	111	110
101	100	102
120	122	121
210	212	211
202	201	200
221	220	222

$x_1 + 2x_2 + x_3 =$		
0	1	2
000	020	010
011	001	021
022	012	002
121	111	101
110	100	120
102	122	112
201	221	211
220	210	200
212	202	222

$x_1 + 2x_2 + 2x_3 =$		
0	1	2
000	002	001
012	020	010
021	011	022
101	100	102
110	112	120
122	121	111
202	201	200
220	210	212
211	222	221

Chapter IV, continued

(6)   No, because any $\mu + 1$ points define a μ-flat only if they are not all in a flat of lower dimension.

(7)

11	21	02	21	01	12	01	11	22
12	22	03	22	02	13	02	12	23
13	23	01	23	03	11	03	13	21
01	02	03	11	12	13	21	22	23

This is isomorphic to the $C(1, 2, 3, 9)$. The correspondence is as follows: 11–1   02–2   21–3   13–7   23–9   22–4   01–5   12–6   03–8

BIBLIOGRAPHY

[1] Albert, A.A., *Fundamental concepts of higher algebra*. Univ. of Chicago Press, Chicago, 1956.

[2] —————, On non-associative division algebras. *Trans. Amer. Math. Soc.*, 72 (1952), 296 - 309.

[3] Atiqullah, M., On a property of balanced designs. *Biometrika*, 48 (1961), 215 - 218.

[4] Baumert, L., Golomb, S.W. and Hall, M., Jr., Discovery of an Hadamard matrix of order 92. *Bull. Amer. Math. Soc.*, 68 (1962), 237 - 238.

[5] Bhattacharya, K.N., A new balanced incomplete block design. *Science and Culture*, 9 (1944), 508.

[6] Bose, R.C., On the construction of balanced incomplete block designs. *Ann. Eugenics*, 9 (1939), 353 - 399.

[7] —————, The affine analogue of Singer's theorem. *J. Ind. Math. Soc.*, 6 (1942), 1 - 15.

[8] —————, A note on the resolvability of balanced incomplete block designs. *Sankhyā*, 6 (1942), 105 - 110.

[9] —————, Mathematical theory of symmetrical factorial designs. *Sankhyā*, 8 (1947), 107 - 166.

[10] —————, On a method of constructing Steiner triple systems. *Contrib. to prob. and statistics: Essays in honor of Harold Hotelling*. Stanford U.P., 1960.

[11] Bose, R.C. and Ray-Chaudhuri, D.K., On a class of error-correcting binary group codes. *Information and Control*, 3 (1960), 68 - 79.

[12] Bruck, R.H. and Ryser, H.J., The non-existence of certain finite projective planes. *Can. J. Math.*, 1 (1949), 88 - 93.

[13] Bussey, W.H., Galois field tables for $p^n \leqslant 169$. *Bull. Amer. Math. Soc.*, 12 (1905), 22 - 38.

[14] —————, Galois field tables of order less than 1000. *Bull. Amer. Math. Soc.*, 16 (1909), 188 - 206.

[15] Carmichael, R.D., *Introduction to the theory of groups of finite order*. Ginn & Co., Boston, 1937.

[16] Chowla, S., A property of biquadratic residues. *Proc. Nat. Acad. Sci. India, A*, **14** (1944), 45-46.

[17] ——————, On difference sets. *Proc. Nat. Acad. Sci. U.S.A.*, **35** (1949), 92-94.

[18] Chowla, S. and Ryser, H.J., Combinatorial problems. *Can. J. Math.*, **2** (1950), 93-99.

[19] de Bruijn, N.G. and Erdös, P., On a combinatorial problem. *Indag. Math.*, **10** (1949), 421-423. Also in *Nederl. Akad. Wetensch. Proc.*, **51** (1948), 1277-1279.

[20] Dickson, L.E., On finite algebras. *Nachr. k. Ges. Wiss. Goettingen., Math.-Phys. Kl.*, (1905), 353-393.

[21] Evans, T.A. and Mann, H.B., On simple difference sets. *Sankhyā*, **11** (1951), 357-364.

[22] Gleason, A.M., Finite Fano planes. *Amer. J. Math.*, **78** (1956), 797-807.

[23] Gordon, B., Mills, W.H. and Welch, L.R., Some new difference sets. *Can. J. Math.*, **14** (1962), 614-625.

[24] Hadamard, J., Résolution d'une question relative aux déterminants. *Bull. Sci. Math.*, (2), Part 1, **17** (1893), 240-246.

[25] Halberstam, H. and Laxton, R.R., On perfect difference sets. *Quart. J. Math. Oxford*, (2), **14** (1963), 86-90.

[26] Hall, M., Jr., Cyclic projective planes. *Duke Math. J.*, **14** (1947), 1079-1090.

[27] ——————, Uniqueness of the projective plane with 57 points. *Proc. Amer. Math. Soc.*, **4** (1953), 912-916.

[28] ——————, Projective planes and related topics. *Cal. Inst. Techn.*, 1954.

[29] ——————, A survey of difference sets. *Proc. Amer. Math. Soc.*, **7** (1956), 975-986.

[30] ——————, Automorphisms of Steiner triple systems. *IBM J. Res. and Dev.*, **4** (1960), 460-472. 2.

[31] ——————, *Survey of numerical analysis*. McGraw-Hill, New York, 1962, 518-542.

[32] Hall, M., Jr. and Connor, W.S., An imbedding theorem for balanced incomplete block designs. *Can. J. Math.*, **6** (1953), 35-41.

[33] Hall, M., Jr. and Ryser, H.J., Cyclic incidence matrices. *Can. J. Math.*, **3** (1951), 495-502.

[34] Hall, M., Jr., Dean, J. and Walker, R.J., Uniqueness of the projective plane of order eight. *Math. Comp.*, **10** (1956), 186-194.

[35] Hall, P., On representations of subsets. *J. London Math. Soc.*, **10** (1935), 26-30.

[36] Hamming, R.W., Error detecting and error correcting codes. *Bell System Techn. J.*, **29** (1950), 147-160.

[37] Hanani, Haim, On quadruple systems. *Can. J. Math.*, **12** (1960), 145-157.

[38] —————, On some tactical configurations. *Can. J. Math.*, **15** (1963), 702-722.

[39] Hartley, H.O., Shrikhande, S.S. and Taylor, W.B., A note on incomplete block designs with row balance. *Ann. Math. Statist.*, **24** (1953), 123-126.

[40] Hessenberg, G., Beweis des Desarguesschen Satzes aus dem Pascalschen. *Math. Ann.*, **61** (1905), 161.

[41] Hilbert, D., *The foundations of geometry* (transl. E.J. Townsend). Kegan Paul & Co., London, 1921 (2nd edn).

[42] Hilbert, D. and Cohn-Vossen, S., *Geometry and the imagination* (transl. P. Nemenyi). Chelsea Publishing Co., New York, 1952.

[43] Isbell, J.R., An inequality for incidence matrices. *Proc. Amer. Math. Soc.*, **10** (1959), 216-218.

[44] Järnefelt, G. and Kustaanheimo, P., An observation on finite geometries. *11-te Skand. Mat. Kongress, Trondheim, 1949*, Oslo, 1952, 166-182.

[45] Kantor, S., Die Configuration $(3,3)_{10}$. *Sitz.-ber. Kais. Akad. Wiss.*, **34** (1881) 2. Abt., 1291-1314.

[46] Khatri, C.G. and Shah, S.M., An inequality for balanced incomplete block designs. *Ann. Inst. Statist. Math.*, **14** (1962), 95-96.

[47] Kirkman, T.P., Note on an unanswered prize question. *Camb. and Dublin Math. J.*, **5** (1850), 255-262.

[48] —————, On the perfect r-partitions of r^2-r+1. *Trans. Hist. Soc. Lancs and Cheshire*, **9** (1857), 127-142.

[49] Lehmer, E., On residue difference sets. *Can. J. Math.*, **5** (1953), 425-432.

[50] Levi, F.W., *Geometrische Konfigurationen*. S. Hirzel, Leipzig, 1929.

114

[51] Majindar, K.N., On incomplete and balanced incomplete block designs. *Proc. Amer. Math. Soc.*, **14** (1963), 223-224.

[52] Majumdar, K.N., On some theorems in combinatorics relating to incomplete block designs. *Ann. Math. Statist.*, **24** (1953), 377-389.

[53] Majumdar, K.N., On combinatorial arrangements. *Proc. Amer. Math. Soc.*, **5** (1954), 662-664.

[54] Mann, H.B., The construction of orthogonal Latin squares. *Ann. Math. Statist.*, **13** (1942), 418-423.

[55] Mikhail, W.F., An inequality for balanced incomplete block designs. *Ann. Math. Statist.*, **31** (1960), 520-522.

[56] Moore, E.H., Concerning triple systems. *Math. Ann.*, **43** (1893), 271-285.

[57] _____, Tactical memoranda. *Amer. J. Math.*, **18** (1896), 264-303.

[58] _____, Concerning the general equation of the seventh and eighth degree. *Math. Ann.*, **51** (1898), 417-444.

[59] Nair, K.R. and Kishen, K., Recent developments in experimental design with special reference to work in India. *I.S.I. Bull.*, **37**, III (1960), 161-177.

[60] Nair, K.R. and Rao, C.R., Incomplete block designs for experiments involving several groups of varieties. *Science and Culture*, **7** (1942), 625.

[61] Netto, E., *Lehrbuch der Kombinatorik*. Teubner, Leipzig, 1901.

[62] Neumann, H., On some finite non-desarguesian planes. *Arch. Math.*, **6** (1954), 36-40.

[63] Newman, Morris, Multipliers of difference sets. *Can. J. Math.*, **15** (1963), 121-124.

[64] Paley, R.E.A.C., On orthogonal matrices. *J. Math. Phys.*, **12** (1932/33), 311-320.

[65] Parker, E.T., Remarks on balanced incomplete block designs. *Proc Amer. Math. Soc.*, **14** (1963), 729-730.

[66] Pickert, G., *Projective Ebenen*. Springer, Berlin, 1955.

[67] Plackett, R.L. and Burman, J.P., The design of optimum multi-factorial experiments. *Biometrika*, **33** (1943-6), 305-325.

[68] Primrose, E.J.F., Quadrics in finite geometry. *Proc. Camb. Phil. Soc.*, **47** (1951), 299-304.

[69] Qvist, B., Some remarks concerning curves of the second degree

in a finite plane. *Ann. Acad. Fenn., Ser. A*, 1, *Math. Phys.*, 134 (1952), 27ff.

[70] Raghavarao, D., On balanced unequal block designs. *Biometrika*, 49 (1962), 561-562.

[71] Rao, C.R., Finite geometries and certain derived results in theory of numbers. *Proc. Nat. Inst. Sci. India*, 11 (1945), 136-149.

[72] Ray-Chaudhuri, D.K., Some results on quadrics in finite projective geometry based on Galois fields. *Can. J. Math.*, 14 (1962), 129-138.

[73] —————, Application of the geometry of quadrics for constructing p.b.i.b. designs. *Ann. Math. Statist.*, 33 (1962), 1175-1186.

[74] Reiss, M., Über eine Steinersche kombinatorische Aufgabe, welche im 45. Band dieses Journals, Seite 181, gestellt worden ist. *Crelle's J. reine und angew. Math.*, 56 (1859), 326-344.

[75] Rouse-Ball, W.W., *Mathematical recreations and essays* (revised by H.S.M. Coxeter). Macmillan & Co., London, 1942.

[76] Roy, J. and Laha, R.G., On partially balanced linked block designs. *Ann. Math. Statist.*, 28 (1957), 488-493.

[77] Roy, P.M., A note on the resolvability of balanced incomplete block designs. *Calcutta Stat. Ass. Bull.*, 4 (1952), 130-132.

[78] Ryser, H.J., A note on a combinatorial problem. *Proc. Amer. Math. Soc.*, 1 (1950), 422-424.

[79] —————, Matrices with integer elements in combinatorial investigations. *Amer. J. Math.*, 74 (1952), 769-773.

[80] —————, Maximal determinants in combinatorial investigations. *Can. J. Math.*, 8 (1956), 245-249.

[81] —————, Combinatorial analysis. *Carus Math. Monogr.*, 14 (1963), 104-107.

[82] Satyarayana, U.V., A note on Mersenne's numbers. *Math. Student*, 28 (1960), 7-8 (1962).

[83] Segre, B., Ovals in a finite projective plane. *Can. J. Math.*, 7 (1955), 414-416.

[84] —————, Sui k-archi nei piani finiti de caratteristica 2.

Revue math. pure et appl., **2** (1957), 289-300.

[85] Seiden, E., A theorem in finite projective geometry and an
      application to statistics. *Proc. Amer. Math. Soc.*,
      1 (1950), 282-286.

[86] Seiden, E., A supplement to Parker's "remarks on balanced
      incomplete block designs". *Proc. Amer. Math. Soc.*,
      14 (1963), 731-732.

[87] Shrikhande, S.S., On the non-existence of certain difference
      sets for incomplete block designs. *Sankhyā*, **11** (1951),
      183-184.

[88] _____, On the non-existence of affine resolvable
      balanced incomplete block designs. *Sankhyā*, **11** (1951),
      185-186.

[89] _____, Relations between certain incomplete block
      designs. *Contrib. to prob. and statistics: Essays in honor
      of Harold Hotelling*, Stanford U.P., 1960.

[90] Singer, J., A theorem in finite projective geometry and some
      applications to number theory. *Trans. Amer. Math. Soc.*,
      43 (1938), 377-385.

[91] Slepian, D., A class of binary signalling alphabets. *Bell
      System Techn. J.*, **35** (1956), 203-234.

[92] Smith, C.A.B. and Hartley, H.O., The construction of Youden
      squares. *J. Royal Statist. Soc. (B)*, **10** (1948), 262-263.

[93] Sprott, D.A., A note on balanced incomplete block designs.
      Can. J. Math., **6** (1954), 341-346.

[94] _____, Balanced incomplete block designs and tactical
      configurations. *Ann. Math. Statist.*, **26** (1955), 752-758.

[95] _____, Some series of balanced incomplete block
      designs. *Sankhyā*, **17** (1956-7), 185-192.

[96] Stanton, R.G., A note on balanced incomplete block designs.
      Ann. Math. Statist., **28** (1957), 1054-1055.

[97] Stanton, R.G. and Sprott, D.A., A family of difference sets.
      Can. J. Math., **10** (1958), 73-77.

[98] Steiner, J., Eine kombinatorische Aufgabe. *Crelle's J. reine
      und angew. Math.*, **45** (1853), 181-182.

[99] Swift, J.D., Construction of Galois Fields of characteristic
      two and irreducible polynomials. *Math. Comp.*, **14** (1960),
      99-103.

[100]   Tallini, G., Le geometrie di Galois e le loro applicazioni alla statistica e alla teoria dell'informazione. *Rendic. Matem.*, **19** (1960), 379 - 400.

[101]   Tarry, G., Le problème des 36 officiers. *C.R. Acad. Franc. pour l'Avancement de Science Naturel*, **1** (1900), 122 - 123; **2** (1901), 170 - 203.

[102]   Vajda, S., *Mathematical Programming*. Addison-Wesley, Reading (Mass.), 1961.

[103]   ————, *The Mathematics of Experimental Design*. Griffin, London, 1967.

[104]   Veblen, O. and Bussey, W.H., Finite projective geometries. *Trans. Amer. Math. Soc.*, **7** (1906), 241 - 259.

[105]   Veblen, O. and Maclagan-Wedderburn, J.H., Non-Desarguesian and non-Pascalian geometries. *Trans. Amer. Math. Soc.*, **8** (1907), 379 - 388.

[106]   Yates, F., Complex experiments. *J. Roy. Statist. Soc.*, *Suppl.*, **2** (1935), 181 - 247.

[107]   ————, Incomplete randomized blocks. *Ann. Eugen.*, **7** (1936), 121 - 140.

[108]   Youden, W.J., Use of incomplete block replications in estimating tobacco-mosaic virus. *Contrib. Boyce Thompson Inst.*, **9** (1937), 41 - 48.

[109]   ————, Linked blocks: a new class of incomplete block designs. (Abstract), *Biometrics*, **7** (1951), 124.

INDEX

order of projective geometry, 28

Pairwise balanced, 70
Pappus, 32
parabola, 45
parallel, 37, 45
partial plane, 45
Pascal, 33
perfect partition, 24
permutation, 5
point, 25, 49
primitive element, 11
projective geometry, 28, 65

Quadratic residue, 21
quadric, 58

Regular group, 6
residual design, 96
resolvable, 89
ruled quadric, 58

Secant of a conic, 42

self-conjugate, 2
simple isomorphism, 4
Singer's theorem, 57, 66
space, 47
Steiner triple system, 73
subgroup, 3
Sylow subgroup, 4
symmetric configuration, 68
— group, 5
system of difference sets, 22, 57
— — distinct representatives, 77

Tangent of a conic, 42
transitive group, 6
treatment structure matrix, 80
type of Abelian group, 5

Unruled quadric, 58

(v,k,λ) configuration, 72
variety, 68

Youden square, 76